Fissile Materials in a Glass, Darkly

Technical and Policy Aspects of the Disposition of Plutonium and Highly Enriched Uranium

Arjun Makhijani, Ph.D.

Annie Makhijani

IEER PRESS

IEER PRESS

Institute for Energy and Environmental Research
6935 Laurel Avenue
Takoma Park, Maryland 20912
Phone (301) 270-5500
Fax: (301) 270-3029

© 1995 Institute for Energy and Environmental Research
January 1995
Second Edition

ISBN 0-9645168-0-2

Library of Congress Catalog Card Number 95-75104

Cover photograph by Robert Del Tredici, Atomic Photographers Guild
Cover and book design by Cutting Edge Graphics
Printed on recycled paper.

Dedicated to
our parents

Bhagwandas and Gopi Makhijani
and
Pierre and Nelly Lepetit

Table of Contents

Preface ix
Summary and Recommendations 1
1. Plutonium and Highly Enriched Uranium as Liabilities 9
   Plutonium 10
   Highly Enriched Uranium (HEU) 16
2. Criteria for Plutonium Disposition 19
   Overview of Disposition Options for Plutonium 20
3. Burning Plutonium in Reactors 24
   Once-Through Fuel Use 25
   Reprocessing and MOX Fuel Use 30
   Conclusion Regarding MOX Fuel 33
4. Plutonium Vitrification: Basic Technical Issues 35
   Processing Plutonium as a Waste 35
   Processing Plutonium for Vitrification 37
   Additives in Plutonium Vitrification 39
   Vitrification of Scrap and Residues 47
   Environmental Controls and Worker Health and Safety 49
   Accidents 50
5. Facilities for Vitrification and Plutonium Processing 51
   Existing Vitrification Plants 51
   Some Practical Considerations Regarding the Use of
      DWPF and WVDP 57
   New Vitrification Plants 60
   Existing Facilities for Processing of Plutonium Metal to
      Prepare it for Vitrification 63
6. Repository Disposal of Plutonium-Containing Waste:
   Some Technical Considerations 65
   Repository Performance 66
   Sub-Seabed Disposal 70

7. Highly Enriched Uranium Disposition: Technical Aspects ... 71
Worldwide HEU Inventory ... 72
Blending Down of HEU for Use in Light Water Reactors ... 73
Blending Down HEU into LEU ... 74
Blending Down of HEU in the United States ... 80
Blending Down in Russia ... 81
Vitrification of HEU ... 81
Hazards of Uranium Processing and Storage ... 82
Environmental Issues ... 83
Diversion of HEU ... 85

8. Policy Issues ... 86
Short- and Medium-Term Issues—Plutonium Disposition ... 86
Highly Enriched Uranium Disposition Policy ... 92
Institutional Issues ... 94
Long-term Policy Issues ... 97

References ... 102

Glossary ... 107

Appendix A. Text of a letter on plutonium sent to President Clinton on October 19, 1994 ... 109

Appendix B. Physical, Nuclear, and Chemical Properties of Plutonium ... 115

Appendix C. Uranium: Its Uses and Hazards ... 121

Figures
1. Steps for plutonium vitrification ... 39
2. Flow sheet for high-level waste solidification at Savannah River Site ... 52
3. Savannah River Site DWPF melter ... 53
4. Glass canister specifications, Savannah River Site ... 54
5. Approaches to blending down HEU ... 79

Preface

Disposing of plutonium and highly enriched uranium (HEU) has become a vexing problem at the end of the Cold War as nuclear weapons are dismantled in large numbers and more and more of these materials become surplus to military requirements even in official terms. Further, non-nuclear weapons states are demanding more insistently that nuclear weapons states take more seriously the spirit of their commitment, under Article VI of the Non-Proliferation Treaty (NPT), to create and implement a timetable for nuclear disarmament. It is therefore possible that all or almost all of these materials, once thought essential for national security by the nuclear weapons states, will become surplus. The possibility that they will be reused in nuclear weapons by the nuclear weapons states themselves or sold illegally for such use by others will continue to pose grave dangers to global security. Effective reduction of nuclear weapons, to say nothing of effective disarmament, will require that these materials be put into a form that cannot be easily reused in nuclear weapons.

This study addresses the disposition of plutonium and highly enriched uranium (HEU) as it relates to putting these weapons-usable materials into non-weapons-usable forms in the short- and medium-term. This is a multi-faceted subject; we consider the following aspects of it:

- Methods of putting weapon-grade plutonium metal and HEU into non-weapons-usable forms;

- Some long-term aspects of plutonium disposition and their compatibility with short- and medium-term actions;

- Disposition of plutonium separated from spent fuel originating in civilian power plants (since this material can also be used to make nuclear weapons);

- Compatibility of fissile materials disposition policies with achieving a halt to all reprocessing;
- Linkages between policies for plutonium and HEU disposition so as to achieve non-proliferation goals effectively.
- Institutional issues related to fissile materials disposition, which also generally apply to storage.

Many important problems associated with plutonium and HEU management are outside the scope of this report, which is in the nature of a monograph. In particular, we do not consider the issues of storage, materials accounting, and safeguards. Plutonium and HEU must be carefully accounted for and safeguarded to prevent their reuse in nuclear weapons or their sale on the black market. There is general agreement on the importance of such safeguards. There is also accord on the importance of safe storage in conformity with protection of health of workers and nearby communities. However, whether this general level of consensus can be translated into specific agreements on the details and into actual practice remains an open question.

For reasons explained in the introduction to this report, we take as a premise that plutonium is an economic, security, and environmental liability. This is surely not a universally accepted idea. The large financial investments that many countries have made in using plutonium as an energy source have created bureaucratic and institutional resistance to new economic and security realities. This inertia has been largely overcome in the United States due to farsighted policies that were initiated in 1976, at the end of the Ford administration, institutionalized by the Carter administration, and carried forward to the present. U.S. policy has therefore long recognized the fact that all plutonium, whether of civilian or military origin, can be used to make nuclear weapons. The use of reactor-grade plutonium in a nuclear weapon was successfully demonstrated in a 1962 test conducted by the United States at the Nevada Test Site, and is a well established fact.

Since the United States has already given up civilian use of plutonium for non-proliferation as well as economic reasons, we believe that it is in an excellent position to exercise global leadership on this crucial issue. Its ability to translate that position into effective action will depend on whether the disposition options that it chooses for its own surplus plutonium take into account the international repercussions of its internal decisions.

As regards highly enriched uranium, most studies assume that it will be mixed with depleted uranium, slightly enriched uranium, or natural uranium in order to convert it into the 3 to 5 percent low enriched uranium (LEU) fuel suitable for use in light water reactors, the most common power reactor design in use today. This report does not examine the economics of this option relative to treating HEU as a waste. Rather, we have pointed out the necessity for examining further options for HEU because, for reasons explained in this report, the option of blending down may not be implemented with the speed necessary to meet growing security concerns.

Like other researchers, we have found that there is no good solution to the disposition of weapons-usable fissile materials; we must select from a menu of poor choices. There are no currently feasible solutions that will get rid of these materials for good. Those that have been proposed as possible options for the future present their own problems of potentially increasing proliferation threats, creating new environmental problems and/or aggravating old ones, and huge costs. Even the exploration of these methods is tied up with unresolved and contentious political questions regarding the future of nuclear arsenals and of nuclear power.

Fissile materials are, in general, necessary for building nuclear explosives. They are defined as materials whose nuclei release energy when split and which can be split with both slow and fast neutrons. Fissile materials in sufficient quantities, called critical masses, can sustain chain reactions and can therefore be used to fuel nuclear reactors. Certain fissile materials, such as natural uranium and low enriched uranium, cannot be used to make nuclear weapons since they cannot be assembled into supercritical masses in which the chain reaction grows so rapidly that there is a large and very sudden energy release—that is, there is an explosion. There are only three weapons-usable fissile materials of practical import to this book, plutonium-239 (in various mixtures), uranium-235 (in HEU), and uranium-233. Uranium-233 does not occur in nature, its man-made stocks are very small relative to plutonium and HEU, and it has not been used in nuclear weapons, so far as public data indicate. We will not consider uranium-233 in this report.

The title of our report draws upon a passage in the Bible that recognizes the uncertainties that are inherent in the human condition whenever we try to peer into the future. Philosophers have generally assigned certitudes to the province of God. The biblical text (from the first Epistle of Paul the Apostle to the Corinthians) reads:

> For we know in part, and we prophesy in part.
>
> But when that which is perfect is come, then that which is in part shall be done away.
>
> When I was a child, I spake as a child, I understood as a child, I thought as a child: but when I became a man, I put away childish things.
>
> For now we see through a glass, darkly; but then face to face: now I know in part; but then I shall I know even as also I am known.

The creation of vast quantities of fissile materials has accentuated all the incertitudes that we are heir to. The present global predicament with respect to weapons-usable fissile materials, whose half-lives are far greater than the longevity of human institutions, has arisen in large measure because governments and their nuclear establishments did not even consider the question of what future generations might do with these materials, if society did not want them. A failure now to recognize the threat to ourselves and to future generations and to deal with it urgently would compound tragically that historic mistake. We must attempt to minimize the risks for our children, even as we recognize the weaknesses of our solutions.

For the purposes of illustrating some of the calculations in this paper, we have taken a notional amount of plutonium (50 metric tons) to illustrate the time frames that would be involved in plutonium vitrification in the U.S. We have not attempted to deal with the problems of exactly how much plutonium may be declared a surplus because, as noted above, this will depend on future arms reduction agreements and on the course and quantity of civilian reprocessing. The amount chosen here, 50 metric tons, is about half of the U.S. military inventory, not including plutonium in un-reprocessed spent fuel.

The portion of this report related to vitrification of plutonium is based partly on a 1992 draft report which IEER prepared for the Office of Technology Assessment of the U.S. Congress (Contract Number I3-4080.0) as a background paper for use in preparation of OTA's own 1993 report, *Dismantling the Bomb and Managing the Materials* (see reference list). However, the present report is IEER's alone, and OTA has no responsibility for its publication or its contents.

Annie Makhijani, co-author of this work and Project Scientist at IEER, researched and wrote most of the chapter on HEU disposition. She also researched many aspects of plutonium chemistry relevant to this report.

Preface

I would like to thank John Plodinec of the Westinghouse Savannah River Company for information regarding vitrification at the Savannah River Site and Ray Richards of Glasstech for information on stirred glass melters. Professor Marvin Miller of the Massachusetts Institute of Technology kindly provided a copy of a recently completed Master of Science thesis by Kory William Budlong Sylvester that contains analyses of important experimental work on and computer modeling of vitrification. Norton Haberman of the DOE and Norman Brandon of Nuclear Fuel Services provided invaluable information on blending down HEU. Charles Forsberg of Oak Ridge National Laboratory provided much information, including data on a new method of vitrifying plutonium that could be especially applicable to plutonium residues.

The National Academy of Sciences study on plutonium, published in 1994, has been invaluable in preparing this work, as the many footnotes referring to it will attest. In this work we have tried to narrow the options further and to integrate disposition of military plutonium, civilian plutonium, plutonium residues, and HEU into a single overall policy.

A number of people provided very valuable review comments that have helped make this a better report. They are: Norman Brandon, Brian Costner of Energy Research Foundation, Charles Forsberg, Beverly Gattis of Serious Texans Against Nuclear Dumping, Ralph Hutchison of Oak Ridge Environmental Peace Alliance, Pete Johnson of the Office of Technology Assessment, J.M. McKibben of the Westinghouse Savannah River Company, Marvin Miller, John Plodinec, IEER's Outreach Coordinator Noah Sachs, and Kathleen Tucker of the Health and Energy Institute. Of course, only the authors of this report are responsible for any errors and omissions in it, and for its contents generally.

The first edition of this report was discussed at IEER's National Symposium on weapons-usable fissile materials held on November 17 and 18, 1994 at the Carnegie Endowment for International Peace in Washington, D.C. The basic technical content of this edition, which is being issued as a book, is the same as that of the first edition, but we have drawn on the suggestions made during the symposium to improve the report and include some new material. Further, as a result of the discussion during the symposium of vulnerability of various forms of glass to theft, we have emphasized one option for the vitrification of plutonium as more desirable than others in this edition. We have added a discussion of how the vitrification program might be carried out rapidly and yet with effective public participation. The background material for this new discussion on contracting and public participation (see Chapter 8) was drafted by Brian Costner. Finally, we have made some editorial changes as a result of further review and the discussion of the work during the symposium. Janna Rolland prepared a summary of the symposium proceedings which was very helpful to the production of the second edition. Tessie Topol helped proofread the manuscript. Todd Perry provided an editorial review and many useful comments.

This report is part of IEER's outreach project on plutonium which is supported by grants from the W. Alton Jones Foundation, the John D. and Catherine T. MacArthur Foundation, and the C.S. Fund, as well as a general support grant from the Public Welfare Foundation.

Arjun Makhijani
Takoma Park, Maryland
January 1995

Summary and Recommendations

Major Findings and Recommendations

Putting plutonium and HEU into forms not easily usable for making nuclear weapons is one of the most urgent security problems facing the world today. A great deal of the urgency derives from the severe economic decline that has occurred in the former Soviet Union since the late 1980s. Several political upheavals have accompanied that decline and the time-scale for these political changes has been on the order of a year or two. Further upheavals are possible and, if economic decline is not reversed soon, likely.

Despite the progress that has occurred between the United States and Russia on many nuclear-weapons-related issues, neither country has a coherent policy for disposition of nuclear materials. Russia is unlikely to act without U.S. leadership and reciprocity, especially given the rising nationalist sentiment that has accompanied economic decline in Russia in the last two to three years. There are already signs that such sentiments may take the form of Russian government policies favoring preserving large stores of weapons-usable fissile materials and nuclear weapons, rather than reducing them.[1] Thus, the U.S. must develop its disposition policy with an eye to its effects in Russia. Given the danger that a global black market in weapons-usable fissile materials originating in Russia may develop, it is imperative that the United States choose a disposition policy and persuade Russia to do the same.

Weapons-usable plutonium also arises from the reprocessing of civilian spent fuel and this must be included in overall disposition policy. The governments of five key countries—Russia, France, Japan, Britain, and India—regard plutonium as a valuable long-term energy resource.

---

1 Lydia Popova, Director, Nuclear Ecology Program, Socio-Ecological Union, Moscow, oral presentation to the IEER National Symposium on Weapons-Usable Fissile Materials held in Washington, D.C. on November 17 and 18, 1994. See also Associated Press wire story on Russian nuclear scientists' views, November 3, 1994.

They continue to operate reprocessing plants to separate plutonium from civilian spent fuel, but their capacity to use plutonium has lagged far behind the rate of its production. As a result, surpluses of civilian plutonium continue to mount, including in Russia. The United States is the only leading country that has wisely rejected the use of civilian plutonium because of its proliferation dangers and its high costs. It is therefore the only country that is in a position to exercise the leadership to persuade other countries to forgo civilian plutonium production, at least for the time being, and to put all separated plutonium into non-weapons-usable forms.

Low uranium prices and an abundant resource base mean that plutonium will not be an economically viable nuclear fuel for many decades (if ever) even for those who regard it as a valuable resource for the long-term. This could provide a basis for attempting to achieve an interim, but universal, halt to civilian and military reprocessing. U.S. disposition policy must be compatible with exercising the leadership to get to this goal. An interim halt to reprocessing would allow time for the energy and security issues associated with plutonium to be negotiated without continuing to separate plutonium in the meantime.

Most studies have advocated that the United States consider the option of turning plutonium into highly radioactive spent fuel by "burning" some of it nuclear reactors as plutonium-uranium mixed oxide (MOX) fuel. Despite some advantages of this approach, it would create an infrastructure for long-term use of plutonium as a fuel in civilian power plants. This is highly undesirable from a non-proliferation standpoint, and has no economic advantages whatsoever.

Appropriate institutional arrangements for managing nuclear-weapons-usable materials for the long-term are needed. The DOE has made great progress on openness at the national level; it created a new office for disposition of nuclear materials in January 1994. It has also boldly taken the lead in rejecting the Advanced Liquid Metal Reactor, which would legitimize plutonium-based fuels, for plutonium disposition, despite pork-barrel pressures to continue funding it. Yet, nuclear weapons spending continues to be very high. This is evidence that the hold of the nuclear weapons makers, which produced conflicts on interest regarding health and environmental issues in the past, continues to be strong, despite the end of the Cold War. It remains to be seen whether the gains of the past few years, and notably of the last two on openness at the national level, can be generalized throughout the weapons complex and sustained. Accomplishing that consolidation is essential to successful implementation of disposition policy.

Our principal recommendations for plutonium disposition are as follows:

- The United States should formally declare excess plutonium a security, economic, and environmental liability, and forswear its reuse in weapons.

- The U.S. should adopt vitrification of plutonium as the strategy for putting plutonium into a non-weapons-usable form. It should forgo all options that involve the use of any reprocessing or reactor technologies for plutonium disposition in order to help promote the objective of an interim, global halt to reprocessing and to discourage the use of plutonium as a fuel in other countries.

- In the next two years, the U.S. should build three or four pilot plants for the vitrification of plutonium so that any technological problems can be cleared up prior to large-scale implementation, and so that the choice of the best vitrification technology can be made on the basis of a technically sound Environmental Impact Statement on vitrification under the National Environmental Protection Act.

- The U.S. should take the initiative in the creation of an international financial guarantee for the re-extraction of plutonium from glass, should it become an economical fuel in the future. It should link this guarantee to achieving an interim, global halt to reprocessing and vitrifying all civilian and all excess military plutonium globally. Appropriate restraints, including public hearings, must be built into this guarantee so that plutonium is not re-extracted without a clear and unequivocal economic justification. On no account should plutonium be re-extracted for use in weapons.

- A reserve of low enriched uranium (LEU) reactor fuel, created by blending down HEU into LEU, should be created so that an alternative to plutonium will be available for decades.

It does not appear at this stage that there are any serious technical hurdles to the implementation of this policy, which is based on combining already commercial technologies. *If this policy is carried out from the beginning with due attention to environmental, health, and safety concerns of workers and the communities near proposed facilities, it should be possible to put all separated civilian and all excess military plutonium into non-weapons-usable form in a decade or less once the political decision is made to do so.*

Other Findings and Recommendations—Plutonium

- There is no satisfactory solution for plutonium disposition that addresses all important security and environmental concerns for all time frames. We must choose from a menu of options that are all partly unsatisfactory in some respects.
- The U.S. should evaluate three options for plutonium vitrification:
  - Vitrification of plutonium mixed with gamma-emitting fission products so that the resulting glass logs meet the spent fuel standard;
  - Vitrification of plutonium mixed with depleted uranium, or some other similar alpha-emitting element;
  - Vitrification of plutonium with a non-radioactive element, such as europium, that would render the extracted mixture unusable for weapons without expensive and difficult processing.
- Vitrification of plutonium alone could also be considered, but it does not appear to present a sufficient barrier to re-extraction by sub-national groups, and therefore is probably unacceptable from a non-proliferation standpoint.
- The "spent fuel standard" for military plutonium disposition—that is, making plutonium as difficult to re-extract as it is from civilian power plant spent fuel—would be the most appropriate one for the short and medium-term if the only concerns were technical ones of re-extraction difficulty and protection against diversion. Such a standard for disposition using vitrification is currently unacceptable to countries that are reprocessing civilian spent fuel because of the very high cost of re-extraction and because vitrification does not extract any energy from plutonium. The security criteria for evaluation of the choice of a disposition policy should include the potential of the policy to contribute to the goal of an interim, global halt to reprocessing and the speed at which all civilian plutonium and excess military plutonium can be put into non-weapons-usable forms.

Summary and Recommendations

- An option with lower re-extraction costs compared to the spent fuel standard but still high enough to pose great challenges to sub-national groups should be explored. Such an option may help to further the goal of achieving a universal, interim halt to reprocessing. If there were financial guarantees for plutonium re-extraction, should it become economical for civilian power production in countries that are now reprocessing civilian spent fuel, these countries may agree to vitrify their plutonium. Vitrification of plutonium with alpha-emitting heavy metals, such as depleted uranium (or other elements with low gamma-emitting properties belonging to a class of elements called actinides), or with certain non-radioactive elements, such as europium or gadolinium, are options that could meet this criterion if there are appropriately high levels of plutonium dilution. The glass so produced should be safeguarded at the same level as plutonium pits or nuclear warheads. (Pits are the metal spheres that form the nuclear triggers of warheads.) This option by itself will not meet the spent fuel standard, especially so far as resistance to diversion is concerned.

- One way of achieving the spent fuel standard and still having a disposition policy that is compatible with policies needed for an interim halt to reprocessing would be to vitrify plutonium with rare earths or actinides first and add a gamma-emitting fission product, such as cesium-137, to the *canister* (instead of adding fission products to the glass). This would provide the same high resistance to theft as spent fuel and also greatly reduce the amount of fission products for achieving it compared to the option of mixing the fission products in the glass itself. As a result, worker exposures and other health and environmental risks may be lower compared to other spent fuel standard options. A feasibility study and laboratory experiments should be initiated to examine this option. *This option appears to be the most promising of all the options that we have examined for achieving the principal disposition goals to the maximum feasible extent.*

- The U.S. should address disposition of plutonium scrap and residues as part of its overall plutonium disposition plan. Because of proliferation concerns, it should rule out all options for processing of residues that, in practice, promote development of reprocessing technologies, such as pyroprocessing. The inclusion of residues in disposition policy will also be very important for non-proliferation and materials accounting in Russia. The U.S. should stop funding the development of pyroprocessing even as a plutonium disposition option. One pilot plant for plutonium vitrification should be devoted to the problem of processing scrap and residues. According to our preliminary evaluation the use of a new technology for direct vitrification of residues, developed by Oak Ridge National Laboratory, appears to be a promising choice for this plant.

- The use of the existing vitrification plants at Savannah River Site, South Carolina and West Valley, New York, presents severe practical difficulties. A feasibility study to examine the use of the plant at Savannah River Site for plutonium vitrification at the time when the melter is scheduled to be replaced should be initiated. The start-up of these two plants for high-level waste vitrification should not be delayed because of their potential use for plutonium vitrification.

- The security problems arising from plutonium cannot be fully resolved even in theory until there is a halt to nuclear power, since nuclear power plants generate plutonium. The U.S. Department of Energy should initiate a fresh evaluation, with full public participation, of the long-term security issues arising from the use and spread of nuclear power plants in light of the severe practical difficulties that have arisen in considering disposition of excess military plutonium.

Other Findings and Recommendations—HEU

- Unlike plutonium, HEU could, in principle, be blended down to provide an economical nuclear power reactor fuel substitute for uranium from mines, so long as there is a market for such fuel.

- The use of LEU made by blending down HEU as a substitute for mined uranium has a number of environmental advantages, such as preventing the accumulation of new radioactive mill tailings and saving energy used in uranium enrichment. These advantages can be realized only if blending down is done in strict conformity with U.S. health and environmental laws. It should be noted that this conclusion assumes that a substantial fraction of existing nuclear power plants will continue to operate for at least the next decade-and-a-half or so. It does not address any environmental or economic issues associated with continuing to run particular nuclear power plants relative to implementing efficiency measures and/or building other types of power plants to replace nuclear capacity.

- The potential re-enrichment of LEU, especially using gas centrifuge technology, will continue to pose proliferation risks even after HEU is blended down.

- The potential advantages of using LEU derived from HEU are partly offset by the realities that U.S. blending down capacity is at present small and verification provisions in Russia to ensure that HEU is actually being blended down are not in place. As a result, storage of HEU in the United States and Russia will continue for a considerable time unless the pace of implementation is increased. Long storage periods increase security risks from potential black-market sales, notably of Russian HEU. Greater capacity for blending down HEU is needed in the near-term if the blending down option is pursued.

- The U.S. should adopt a policy of reciprocity vis-à-vis Russia as regards HEU disposition policy. This will lead to a more equitable and secure reduction of a larger portion of global HEU stocks than planned under the current U.S.-Russian agreement.

- Vitrification of HEU as an interim measure may reduce security risks arising from the potential for black market sales. However, it may also make the use of future use of LEU derived from vitrified HEU uneconomical relative to LEU from mined uranium. Prior to choosing an option, the DOE should include in its Programmatic Environmental Impact Statement on disposition of fissile materials a careful evaluation, with full public participation, of the security and environmental concerns of extended HEU storage compared to conversion of larger amounts of HEU to LEU and vitrification of much or most HEU.

Canisters for vitrified high-level waste, Savannah River Site, South Carolina. (Photo by Robert Del Tredici, Atomic Photographers Guild)

CHAPTER 1

Plutonium and Highly Enriched Uranium as Liabilities

The existence of this surplus material [plutonium and highly enriched uranium] constitutes a clear and present danger to national and international security. None of the options yet identified for managing this material can eliminate this danger; all they can do is to reduce the risks.

National Academy of Sciences' 1994 report on plutonium[2]

With the end of the Cold War, weapons-usable fissile materials have emerged as one of the most important security threats to the world. Surpluses of plutonium and highly enriched uranium have arisen from the dismantling of unwanted nuclear warheads. As the Soviet Union disintegrated in the early 1990s, and as the Cold War arrangements of influencing smaller countries in the world gave way to uncertainty, the possibility has increased that some of these surpluses (or even the nuclear warheads themselves) may be sold illegally, with unpredictable human, military, political, and environmental consequences. In a crisis, Russia or the U.S. could reuse some of these fissile materials from dismantled weapons to make new warheads. This would likely result in a similar response from the other side. Therefore, ready availability of weapons-usable fissile materials would make it easier and faster for one side to reignite the arms race. It also makes non-proliferation policy less effective, since non-nuclear-weapons states are less likely to believe that

---

2   NAS 1994, p. 1.

surplus plutonium and HEU will not again be used in weapons if these materials remain in weapons-usable forms.

Plutonium

Plutonium is made by the irradiation with neutrons of uranium-238 in military as well as civilian nuclear reactors.[3] In order to be used in weapons, plutonium must first be separated from un-used uranium and from fission products in the reactor fuel and target rods. This chemical separation process, known as reprocessing, is one of two key technologies in the production of nuclear-weapons-usable fissile materials. (The other technology is uranium enrichment—see below.)

Plutonium from civilian reactors as well as that from military reactors can be used for making nuclear weapons. There are some important differences between the characteristics of plutonium produced in the most common civilian reactors (light water reactors) and military plutonium. The former, known as "reactor grade plutonium," has a larger proportion of plutonium isotopes other than plutonium-239, the one most suitable for weapons. These other isotopes, notably plutonium-240 and plutonium-241 (as well as americium-241, which is the decay product of plutonium-241), make it somewhat more complex to make a nuclear weapon of predictable yield from reactor grade plutonium, whose use also entails larger radiation doses to workers. Neither of these factors is an effective obstacle to the proliferation problems posed by separated plutonium of civilian origin.

Reactor-grade plutonium has 19 percent or more of plutonium-240, and typically contains 55 to 60 percent plutonium-239. Weapon-grade plutonium has 7 percent or less of plutonium-240, with almost all the rest being plutonium-239. Appendix B shows some important nuclear, physical, and chemical properties of plutonium.

Table 1 shows approximate estimates of the global stocks of plutonium, separated as well as unseparated from irradiated fuel rods, as of 1990.

---

3. Some reactors are dual-use reactors, which generate power for civilian use as well as plutonium for military purposes. Many military reactors have special target rods of depleted uranium which are irradiated to yield weapon-grade plutonium.

TABLE 1. **Global plutonium inventories in 1990**

TYPE OF PLUTONIUM	METRIC TONS
Military plutonium	248
Civilian plutonium, separated	122
Plutonium in civilian spent fuel, unseparated	532

Source: For U.S. military plutonium, Grumbly 1994; for all other data, Albright et al. 1993, p. 197. For this table, Albright et al.'s estimate of U.S. military plutonium of 112.2 metric tons (pp. 34–35) was subtracted from their global total and replaced with the official DOE production figure of 103.5 metric tons.

Note: These estimates are being refined as more recent data are analyzed.

Of the 248 metric tons of military plutonium, almost 200 metric tons will become surplus under current plans to reduce U.S. and Russian arsenals to about 5,000 warheads each, if the term "surplus" (or "excess") is defined as the plutonium not actually in warheads.[4]

The global surplus of plutonium is being increased by separation of plutonium from civilian nuclear power reactor spent fuel. The global cumulative amount of such plutonium through the end of 1980 in all countries was estimated to be about 39 metric tons; it increased about three-fold to about 122 metric tons by the end of 1990. During the same period, a number of countries abandoned or drastically scaled down breeder reactor programs designed to use much of this plutonium, mainly because these programs could not be justified economically. Only about 50 metric tons of this separated plutonium had been used in reactors by 1990; some of that was sitting in the cores of shut-down breeder reactors, and hence was not actually being used. The surplus of civilian plutonium is projected to greatly increase if reprocessing is not drastically curtailed.[5]

---

4 This estimate is based on the following assumptions: (i) there are, on average, under 4 kilograms of plutonium in each warhead and (ii) there are about 20 metric tons of plutonium in the military inventories of other nuclear weapons powers. To maintain an arsenal at a given size, an additional small inventory (relative to the plutonium in the weapons) is required in order to compensate for accidental or remanufacturing losses. David Wright of the Union of Concerned Scientists calculates that this would amount to less than one metric ton for the projected U.S. arsenal (Wright as cited in IEER 1994, p. 2). This amount is much smaller than the uncertainties in the above calculations and so can be ignored in the present context.

5 Albright et al. 1993, pp. 204–207.

It has also become clear in the last two decades that economically recoverable world resources of uranium are much larger than estimates made in the 1950s and 1960s, when plutonium separation was deemed by many to be essential to the future of nuclear energy. In the past few years a number of analyses in the United States have convincingly demonstrated that plutonium is not economical as an energy source and will not be for the foreseeable future because of the high costs of breeder reactors, reprocessing, and fabrication of fuel containing plutonium.

These analyses have examined the least expensive of the options for using plutonium for electricity production. This involves converting plutonium into plutonium dioxide, mixing it with uranium dioxide (the fuel form used in the most common nuclear power reactor design in the world today, the light water reactor) to obtain "mixed oxide" fuel (abbreviated as MOX fuel). The costs of plutonium processing are so high that even if the separated plutonium is considered free, a reasonable assumption for surplus plutonium from unwanted nuclear warheads, uranium is still cheaper as a nuclear power plant fuel. John H. Gibbons, President Clinton's Assistant for Science and Technology, summed it up succinctly in Congressional testimony in May 1994: "Contrary to some claims, there is no money in plutonium—except, perhaps on the nuclear black market."[6]

We will not repeat the analyses that have already been made in previous studies, notably the 1994 study on plutonium disposition by the National Academy of Sciences,[7] a 1993 analysis of fissile materials by the RAND Corporation,[8] and a 1992 study by Berkhout and his colleagues at the Center for Energy and Environmental Studies at Princeton University.[9] The basic conclusion regarding the economics of nuclear reactor fuel is very clear. The prevailing spot price of uranium oxide (yellowcake) is well below $10 per pound.[10] According to the RAND analysis, if the cost of reprocessing is taken to be equal to the charges for reprocessing of about $1,600 per kilogram of heavy metal (approximately equal to the combined uranium and plutonium content of the spent fuel), and the yellowcake price is assumed to be $10 per pound, MOX fuel

---

6   Gibbons 1994, p. 3.
7   NAS 1994
8   Chow and Solomon 1993.
9   Berkhout et al. 1992.
10 Wall Street Journal, October 24, 1994. The price is rising toward $10 per pound, perhaps more. We assume, for the purposes of this report, a yellowcake price of $10 per pound. It is expected to be in the $7 to $15 range over the next few years.

would not be competitive until yellowcake prices increased about 16-fold to $160 per pound. Further, according to the same analysis, even if the capital cost of the reprocessing plant is ignored, MOX fuel would not be competitive until uranium prices quintupled.[11] The RAND report's conclusions are similar to those in the earlier analysis by Berkhout et al.[12]

The prospects that plutonium will ever be an economical energy source are very slim. However, proponents of civilian plutonium use in countries such as Japan and France, which do not have large domestic supplies of fossil fuel resources, have argued that development of the technology for plutonium use is essential for the very long-term future; they claim that there are no viable alternatives to plutonium on the scale of energy supplies that they are likely to require. Such arguments are especially forceful in Japan, which does not appear to have ample domestic uranium resources and where the land area for potential development of solar energy is very limited.

The modest theoretical merit of such arguments is overwhelmed by a number of realities. First, the danger of plutonium diversion is very real, especially in the context of continued economic, political and military instability and uncertainty in the former Soviet Union. Continued arguments that some countries need plutonium separation now for potential use in some distant future only encourages further plutonium separation and development of ancillary facilities in Russia.

The risk of diversion exists in all countries that own plutonium, though it is now most acute in Russia. The large-scale use of plutonium in the civilian sector will create new opportunities for diversion and for involvement of organized criminal elements in the traffic. Finally, the use of civilian plutonium in Western Europe and Japan creates obstacles to the stopping of reprocessing in Russia by depriving the United States of important leverage in dealing with Russia. The U.S. can hardly turn a blind eye to reprocessing in Western Europe and Japan while persuading Russia to stop.

Second, the security benefits of rapidly vitrifying separated plutonium are great and incalculable, while the costs of vitrifying plutonium, especially if it is done without mixing fission products in the glass, are

---

11 Chow and Solomon 1993, pp. 35–39. There are various estimates of reprocessing costs. Chow and Solomon state that the $1,600 per kilogram of heavy metal is the mid-point of the range of $1,400 to $1,800 in reprocessing charges actually paid to France and Britain by Japan and other customers, as of the time of the RAND study.

12 Berkhout et al. 1992, pp. 18–20.

relatively modest. The technology for re-extracting plutonium from glass is known, should plutonium ever become an economical fuel. There is therefore no need to continue to operate reprocessing plants to produce more plutonium that is uneconomical today and will remain so for decades, at least. The lead-time needed for construction of re-extraction facilities, should such facilities ever be necessary, is far shorter than any reasonable projected time in which plutonium may become economical as a fuel. The self-sufficiency argument therefore has essentially no merit in the near- and medium-term, since plutonium use cannot contribute to self-sufficiency in this time-frame. Japan will continue to be dependent on both imported oil and uranium. This reality has prompted a proposal that Japan should stockpile uranium instead of plutonium since uranium is plentifully available at low prices.[13]

More broadly, the self-sufficiency argument is rather weak. It received a strong impetus in many countries, including France and Japan, from the sudden increase in oil prices during 1973–1974 and from the embargo imposed by Arab oil exporting countries against the U.S. in late 1973. Many analysts incorrectly believed that exportable oil supplies could be monopolized by a few countries. Since oil was a vital commodity at risk of being cut-off, the argument went, self-sufficiency, or something near to it, was a security and economic imperative.

However, oil, like uranium, has turned out to be far more plentiful than presumed by the self-sufficiency analysis. Natural gas is also more abundant than once thought. There are far more oil exporting countries in 1994 than there were 20 years ago. The increases in the price of oil in the 1973–1980 period were not related to a physical dearth of supply, but to control of exportable supplies by a few countries, which could not be sustained.

If there is an argument for self-sufficiency in energy, it should apply with greater force to food, especially so far as Japan is concerned. Japan imports most of its food supply since domestic food production cannot provide for its present consumption level and pattern. Moreover, Japan has not experienced an oil cut-off, but it has seen one imposed on an essential foodgrain. In 1973, a few months before the Arab oil embargo against the U.S., President Nixon briefly banned all exports of soybeans as part of his program to curb the sudden price increases of commodities and to control inflation. Yet Japan did not set itself the goal of self-sufficiency in food, even though its closest military ally did not prove to

---

13 Leventhal and Dolley 1994.

be a fully reliable supplier of food grain. Rather, it diversified its sources of supply, largely by importing more soybeans from Brazil.

Japan could not sustain anything near its present level of use of resources without continuing to import many other essential commodities. High exports are the necessary counterpart to high imports. In the context of this economic reality, energy independence is an exaggerated and obsolete policy response. Whatever modest merit there might be in energy independence arguments made in Japan and France in support of plutonium separation is far outweighed by the negative security consequences of reprocessing, even if all adverse economic and environmental factors are ignored.

Russia has even less reason to stick with civilian plutonium production because it has huge reserves of various forms of energy, including fossil fuels and uranium. There is also immense room for improving energy efficiency in Russia. Further, Russia has been the scene of the worst civilian and military accidents of the nuclear era, namely the fire in one of the reactors at Chernobyl in 1986 and an explosion in a high-level radioactive waste tank at the Chelyabinsk-65 nuclear weapons plant in 1957. The frequency of accidents in recent years as well as the past record of despoliation of the environment are further reasons for Russia to reconsider its nuclear policies; many people in Russia are working toward that end. Britain also has plentiful fossil fuel reserves, and is an oil exporter.[14]

In contrast to a distant theoretical possibility that plutonium may one day be an economical energy source is the real evidence of a developing black market in fissile materials, including plutonium. The most serious confirmed incident involved an attempt to smuggle about 350 grams of plutonium into Germany; this is not enough plutonium for a nuclear warhead, but more than enough for a radiation dispersal weapon. It is possible that this sale of black market plutonium, originating to all appearances in the former Soviet Union, was in response to a demand created by German secret police to learn more about the potential supply situation. What has been learned is alarming. This incident has shown that plutonium availability depends on the demand for it and indicates that other countries or groups wanting to purchase plutonium could also similarly acquire it. Unlike the German government (which has a large

---

14 This discussion addresses only security aspects of "energy independence"; we do not consider other aspects such as environmental problems and risks associated with various energy sources.

stock of separated plutonium), groups or countries wanting to acquire plutonium for clandestinely building nuclear warheads or radiation dispersal weapons would hardly advertise their successes. In fact, there is no way for the world to know whether any plutonium and highly enriched uranium have already been sold, and if so, how much and to whom. There are still no adequate materials accounts of Soviet production of these materials. Nor are there any transparency and safeguards arrangements in place that would allow a determination of the quantities and flows of the materials. The progress on putting such measures into place has been very limited and far short of the need.

Finally, there is the potential that one or more of the many non-nuclear weapons states that are signatories to the Non-Proliferation Treaty (NPT) and that own separated plutonium could change their minds, and either openly or clandestinely make nuclear weapons. Indeed, the very fact of this potential is an incitement to proliferation, because it increases the level of suspicion between countries. The most notable example is the tension between North Korea and Japan regarding nuclear weapons. North Korea, pointing to Japan's imperialist past, claims that Japan may well make and use nuclear weapons, and that it possesses the technical capability and materials to do so. North Korea's failure to comply with inspection demands by the International Atomic Energy Agency (IAEA) has in turn tentatively raised questions in Japan regarding a potential Japanese nuclear deterrent. These military and political tensions, arising partly from plutonium production in both North Korea and Japan, should be an additional powerful consideration against continued plutonium production and for creating and implementing a policy for disposition of already separated plutonium.

Highly Enriched Uranium (HEU)

As U.S. and Russian nuclear arsenals are reduced, large amounts of HEU, the other fissile material that can be used to make nuclear weapons, are also becoming surplus to weapons requirements, along with military plutonium. While both HEU and plutonium are weapons-usable materials, there are some differences between them. HEU is generally not used in civilian power reactors.[15] Another contrast to plutonium is that HEU is not made in nuclear reactors.

---

15 An exception to this is when HEU is loaded into breeder reactor cores as a substitute for plutonium. HEU is also used as a fuel in some naval propulsion reactors and in some research reactors.

HEU is a special mixture of isotopes of uranium that is made by increasing the uranium-235 content of natural uranium by a process called "enrichment." Natural uranium contains only 0.711 percent uranium-235, the fissile isotope of uranium. Almost all the rest is uranium-238, which is not fissile, though it is the raw material for the production of plutonium-239, which is fissile.[16] The process that is used to make enriched uranium also creates a waste stream of depleted uranium, so called because it contains less fissile uranium-235 than natural uranium.

Uranium must be enriched to high levels of uranium-235 content in order to be usable for making nuclear weapons. At levels of 3 and 5 percent enrichment, it is called low enriched uranium (LEU), which cannot be used to make a nuclear weapon. It must be further enriched in order to make possible the assembly of the super-critical mass required for an explosion. However, LEU is the most common fuel used for the generation of electricity in nuclear power plants. (Some power plants, notably the heavy-water-moderated reactors of Canadian design, use natural uranium and do not require enrichment facilities.)

Weapon-grade enriched uranium typically contains over 90 percent uranium-235. The amount of weapon-grade uranium required for the manufacture of a bomb is about 15 to 20 kg. But weapons can be made with far lower enrichment levels. At 20 percent enrichment, it would take 250 kg to make an explosive device.[17] This may be considered a kind of practical lower limit to the enrichment required for making weapons. Appendix C provides additional information on the properties of uranium. There are about 2,300 metric tons of HEU in the world; almost all of this inventory is in the former Soviet Union and the United States (see Chapter 7).

The process of enrichment of natural uranium can be reversed. To do this, HEU is blended with natural uranium, depleted uranium (which contains 0.2 to 0.3 percent uranium-235), or slightly enriched uranium (0.8 to 2 percent uranium-235), to make LEU for use as power reactor fuel. Leaving aside for the moment the desirability of pursuing such a course, reactor fuel made in this way could, in principle, be competitive with fuel made from uranium ore. Thus, given the existence of reactors that can use LEU fuel as well as of fuel fabrication facilities, HEU is not an economic liability in the same way that plutonium is.

---

16 While uranium-238 is not fissile and cannot sustain a chain reaction, it can be fissioned with fast neutrons to yield energy. This property of uranium-238 is used in advanced nuclear weapons to provide a significant portion of their yield.

17 Chow and Solomon 1993, p. 5.

However, we should bear in mind that LEU can be re-enriched to make HEU. The difficulty of detection of re-enrichment partly depends on the type of equipment used for enrichment. Gas centrifuge technology, which is used commercially to make LEU for power reactors in both Europe and Russia, could be used with relative ease to make quantities of HEU sufficient for one or more nuclear weapons.[18] A privately-owned gas centrifuge plant has been proposed to be built in Louisiana, United States. A license application is pending before the Nuclear Regulatory Commission.

The criteria for selecting disposition options for plutonium and for HEU are similar in that they both represent security threats, but they differ in that the economic and environmental issues associated with their disposition are somewhat different. We will consider plutonium in the next part of this report (Chapters 3 through 6), and then briefly consider issues related to HEU (Chapter 7).

---

18 The quantity of HEU required for a nuclear weapon is about 3 to 4 times greater than that for weapon-grade plutonium. About 3 to 5 kilograms of weapon-grade plutonium are required for a fission weapon, though a recent report by the Natural Resources Defense Council states that a kiloton-range weapon can be made with as little as one kilogram. Weapon-grade plutonium contains about 93 percent of the fissile isotope plutonium-239.

CHAPTER 2

Criteria for Plutonium Disposition

The recognition that plutonium has no practical economic value means that in all cases plutonium disposition will require a net expenditure of funds. This is so even in the options that generate revenues from the sale of electricity obtained from the use of plutonium as a fuel. The efficacy of various options for plutonium disposition can be evaluated according to the following criteria:

1. *Security aspects*: The treatment, storage, and disposal of plutonium as a waste must be such that the difficulty of plutonium re-extraction from the waste is as close to new plutonium production and separation as possible.

2. *Time frame*: Putting plutonium into non-weapons-usable form as soon as possible (compatible with protection of the environment and of worker and community health) is crucial in light of the situation in the former Soviet Union. Russia is unlikely to act without the U.S. doing so also.

3. *Accident risks*: The risk of catastrophic accidents, resulting in the dispersal of plutonium or accidental nuclear or non-nuclear explosions, must be evaluated for each option.

4. *Health, environmental protection, and safety*: The option chosen should be compatible with compliance with all applicable environmental, health and safety laws and regulations. It should take account of the reality that increased handling, processing, and transportation entail additional new environmental risks, and that some of these new risks may offset existing risks from storage.

5. *Potential for encouraging plutonium production*: Some disposition options involve the use of reprocessing technologies and/or of facilities to fabricate fuel containing plutonium. Hence there is a need to consider the potential for a U.S. choice of a disposition option to entrench the separation and use of plutonium in other countries.

6. *Cost*: It is important to compare the costs of various disposal options for plutonium, though in light of the immense security risks involved, this is a secondary issue.

No set of policies designed to deal with plutonium disposition will achieve all these objectives to the greatest possible degree simultaneously. For instance, achieving a high degree of difficulty in re-extraction or even transmuting all plutonium into fission products could be in serious conflict with the objectives of putting plutonium into a form unusable for weapons as rapidly as possible.

Overview of Disposition Options for Plutonium

The 1994 National Academy of Sciences study on plutonium (referred to below simply as the NAS study or the 1994 NAS study) categorized the many options for dealing with plutonium into three groups:[19]

- "Indefinite Storage" of plutonium;
- "Minimized accessibility" to reduce the potential of plutonium being made into weapons;
- "Elimination" of plutonium by methods such as transmutation, which would completely (or nearly so) convert plutonium into fission products.

Under the last two categories, the NAS considered whether the plutonium would be used in reactors or whether it would be disposed of without such use.

As is evident from the term, "indefinite storage" means that "the plutonium would continue to be stored in weapons-usable form indefinitely."[20] While temporary storage is a practical necessity in all cases until plutonium can be put into a more proliferation-resistant form,

---

19 NAS 1994, pp. 144–146.
20 NAS 1994, p. 144.

indefinite storage does not meet the minimum criteria for achieving security goals of preventing black market sales or reuse in weapons. We will not consider this option any further in this report.

The NAS report discusses a large number of options under the second category of "minimized accessibility." Specific criteria related to "accessibility" are needed in order to enable an evaluation and comparison of these options. Like most studies on this subject, the NAS study adopted the "spent fuel standard" as an approximate measure of how inaccessible the plutonium has been rendered to prevent its future use in weapons.

The "spent fuel standard" does not mean that the problem of plutonium is solved; only that it will be approximately as difficult to re-extract and use plutonium for making weapons as it would be to get it by reprocessing civilian spent fuel.

Such a "standard" suggests itself from a practical reality—most plutonium today is not in nuclear weapons or stored pits, but is rather in spent fuel from nuclear power plants. Therefore, the problem of plutonium and proliferation is bound up with the existence of this larger stock of plutonium, and it makes little sense to subject plutonium from weapons to a more stringent non-proliferation standard than spent fuel.

The fact that plutonium in spent fuel is mixed with uranium and with fission products, many of which emit intense gamma radiation, has two consequences of importance to disposition. First, as a result of this external gamma radiation, spent fuel is extremely dangerous to handle—in fact it must be heavily shielded or handled remotely. Any proximity to unshielded spent fuel would result in a lethal dose of radioactivity in minutes (or even less for fresh spent fuel). Second, for the plutonium in spent fuel to be used for nuclear weapons, the spent fuel would have to be reprocessed, a difficult and costly undertaking.

These two characteristics make spent fuel very proliferation-resistant both from the point of view of the potential for theft and the difficulty of re-extraction. However, neither factor prevents countries that have spent fuel from deciding to extract the plutonium present in it. Therefore, the NAS also recommended some research on long-term means to get rid of plutonium altogether, using technologies that would fission all of it. But the spent fuel standard has a serious potential drawback in that it may make it more difficult to achieve a halt to civilian reprocessing and to put separated civilian plutonium into non-weapons-usable forms. We will discuss this issue further in Chapter 8 on policy.

Most options that would minimize accessibility of plutonium for use in nuclear warheads or radiation dispersal weapons fail on one or more of the criteria listed at the beginning of this chapter. We list them in Table 2 and indicate the main reasons for doing so.

TABLE 2. Rejected minimized accessibility plutonium disposition options

DISPOSITION OPTION	PRINCIPLE REASON FOR REJECTION
New burner reactors—No reprocessing	Long-time frame; licensing uncertainties.
New thermal reactors with reprocessing	Encourages reprocessing and hence undermines non-proliferation goals; long time-frame.
Advanced Liquid Metal Reactor (ALMR)	ALMR can be used to breed plutonium; most proposals for its use also require a new reprocessing technology (pyroprocessing); long-time frame; undermines non-proliferation goals.
Pyroprocessing without ALMR	Promotes development of a new reprocessing technology under the guise of plutonium disposition; undermines non-proliferation goals.
Nuclear explosion in an underground cavity	Extensive and unacceptable environmental damage; undermines the non-proliferation goal of stopping nuclear explosions.
Sub-critical reactor with proton accelerator	Involves development of a reprocessing technology and hence undermines non-proliferation goals; long-time frame; high technical uncertainty.

We refer the reader to the NAS study for further discussion of these options. In this report we will consider in more detail three options for minimized accessibility:

- The use of mixed plutonium-uranium oxide fuel in reactors;
- Vitrification of plutonium;

- Repository, deep-borehole, or sub-seabed disposal.

For several reasons we have placed the greatest emphasis on vitrification:

- We have concluded that the choice of MOX fuel use as a U.S. disposition option would be a serious hindrance to achieving an interim global halt to reprocessing and to non-proliferation and disarmament goals that are commitments under the NPT.
- There is an extensive literature on disposing of plutonium as MOX fuel.[21]
- Vitrification is central to our recommendations on plutonium disposition.
- Some vitrification options have not yet been debated and explored in published policy literature to the extent that we feel is warranted by their relative merit.

We will also discuss long-term plutonium disposition issues, since none of the options for minimizing accessibility actually get rid of all the plutonium.

---

21 OTA 1993; Chow and Solomon 1993; NAS 1994; Berkhout et al. 1992.

CHAPTER 3

Burning Plutonium in Reactors

There are a number of proposals for burning plutonium in existing reactors in order to make it unusable in weapons without extensive chemical and physical processing. The plutonium would be rendered unusable for weapons because it would be mixed with large amounts of natural or depleted uranium to turn it into a suitable fuel, and, after use in the reactor, the remaining plutonium would also be mixed with highly radioactive fission products. In order to be used in weapons again, the plutonium would have to be separated both from the fission products and the uranium.

It should be noted that the destruction of plutonium in reactors is rather slow. The reason is that even as some of the plutonium loaded as fuel is being consumed in the fission process, more plutonium is being created from uranium-238 present in the fuel. Spent fuel rods from reactors contain a significant amount of plutonium that can be recovered by reprocessing.

Using plutonium as a power reactor fuel can happen within two very different contexts. One context is that of once-through fuel use, where the unused uranium and plutonium are not re-extracted (by some form of reprocessing) for reuse. Once-through use in existing reactors has been proposed for plutonium disposition in the United States, where there are no reprocessing facilities for civilian spent fuel and where all military reprocessing facilities are shut down. The other context, burning in reactors with reprocessing, prevails in countries such as France which are operating reprocessing plants for civilian spent fuel.

There is also the question of whether new or existing reactors should be used. Two different recommendations have emerged in this regard from recent authoritative studies. The Office of Technology Assessment, in its 1993 report, *Dismantling the Bomb and Managing the Materials*, concluded that "The use of surplus plutonium from weapons as fuel for U.S. commercial reactors is unlikely because of economic factors, the concerns of U.S. utilities about regulatory constraints and public acceptance, and the need to evaluate U.S. policies that discourage commercial plutonium use."[22] As a result of this finding, OTA recommended consideration of a dedicated, government-built and -owned reactor for the purpose of plutonium disposition as one of the two near-term, technically available options. The other option recommended for consideration was plutonium vitrification.[23]

The NAS study, on the other hand, concluded that it would be preferable to use existing reactors, possibly government-owned, if licensing and other concerns such as those cited by the OTA report for private reactors turn out to be difficult issues. It also concluded that new reactors should be considered only if concerns such as public acceptance and licensing turn out to be insurmountable. Even in such a case, "the use of advanced reactors and fuels to achieve high plutonium consumption without reprocessing is not worthwhile, because the consumption fractions that can be achieved—between 50 and 80 percent—are not sufficient to greatly alter the security risks posed by the material remaining in the spent fuel."[24] The NAS was also concerned about the long time period needed to implement plutonium disposition using new reactors. Thus, the NAS report discouraged the consideration of new reactors, and especially new reactor designs, such as the Modular High Temperature Gas-cooled Reactor (MHTGR), for plutonium disposition.[25]

The debate over new reactors (of existing design or close to it) versus existing reactors is a difficult one. We will first discuss general issues associated with once-through MOX fuel use, which would apply in both cases, and then compare the merits of using existing commercial reactors to those of building a new dedicated reactor. Proposals for burning MOX fuel in existing foreign commercial reactors have also been put forward. We will discuss these briefly.

Once-Through Fuel Use

The United States is committed to once-through fuel use in civilian power applications; spent fuel is legally designated as high-level radio-

22 OTA 1993, p. 4.
23 OTA 1993, p. 4.
24 NAS 1994, p. 15.
25 NAS 1994, pp. 183-187.

active waste to be disposed of in a repository, without reprocessing. The adoption of a once-through MOX fuel use for plutonium disposition would be consistent with existing policy. Technically, the approach automatically ensures that disposition meets the spent fuel standard, assuming the fuel is kept in the reactor about the same length of time as uranium fuel. This is the single most important advantage of the once-through plutonium disposition strategy.

Another important advantage of the MOX strategy is that it is a technically proven one. MOX fuel has been used in light water reactors for some time, though the experience is mainly in Europe. Reactors that use LEU fuel can have one-third of their fuel consist of MOX while the rest is conventional, low enriched uranium dioxide. There are also three reactors in the United States at the Palo Verde Nuclear Generating Station in Arizona, known as System-80 pressurized light water reactors (PWRs), that are capable of burning a full MOX core.

Plutonium concentrations as high as six to seven percent may be possible with the use of special neutron absorbing materials. According to the NAS study, this "would require safety review."[26] The plutonium would be mixed with depleted uranium so that essentially all the fissile material would consist of plutonium isotopes. The DOE commissioned a European multinational corporation, ABB, to do an extensive study of the use of these reactors, which was completed in 1993.[27] The NAS panel estimated that with the highest possible concentrations of plutonium it would take 30 reactor-years, or three reactors of the System-80 type operating for 10 years each, to convert 50 metric tons of weapon-grade plutonium into spent fuel.[28]

The time needed to convert 50 metric tons of weapon-grade plutonium into spent fuel would depend on the number of reactors that are employed. In theory, the larger the number of reactors, the faster the conversion to spent fuel. However, this theoretical advantage must be balanced against the multiplication of security risks that arise when the number of sites that handle and use plutonium is increased. Further, the delays in licensing reactors for MOX fuel use are likely to increase (possibly disproportionately) with the number of reactors that use MOX fuel, partly because the age and operating records of reactors and nuclear utilities vary considerably.

26 NAS 1994, p. 159.
27 ABB 1993.
28 NAS 1994, p. 159.

The time requirement for completing the conversion of 50 metric tons of plutonium to spent fuel will be increased not only by licensing requirements and any modifications that might have to be made to existing reactors to burn MOX fuel, but also by the lack of sufficient capacity to make MOX fuel in the first place. The only facility in the United States that could do the job is the Fuel and Materials Examination Facility at Hanford, which would have to be modified. The NAS study cited cost estimates for a new MOX fuel fabrication facility of between $400 million and $1.2 billion and time estimates of a decade or more to build it.[29] Overall, it may take two decades or more to convert 50 metric tons of weapon-grade plutonium into spent fuel.[30]

One important disadvantage of using the MOX fuel option in the United States for disposing of excess plutonium is that it would create an infrastructure for long-term use of plutonium in civilian power plants. The huge sunk costs of such an infrastructure, including the licensing and other requirements of MOX fuel transport and use in reactors, will tend to create financial and political vested interests in continued use of plutonium in commercial power generation. Having created these vested interests in MOX fuel use domestically, the U.S. would find it politically difficult or impossible to oppose the creation of a similar infrastructure in Russia.

The Russian desire for reciprocity on the part of the U.S. is strong, and may become stronger if nationalist sentiment continues to increase in Russia. A U.S. decision to choose the MOX option as its plutonium disposition strategy could encourage the construction of MOX facilities in Russia. The adverse safeguards implications of such facilities would be great. Already, the economic and political stresses in the former Soviet Union are a principal cause of security concerns arising from surplus plutonium. Materials accounting questions in the former Soviet Union are even more severe than they are in the United States. The operation of MOX fuel facilities in Russia would aggravate all these problems.

A MOX fuel fabrication plant is being considered in Russia, and Siemens, the German multinational company, would be a likely partner in the enterprise, if it is built. There is a partially complete MOX fuel plant at Chelyabinsk-65. Construction was halted due to lack of funds.

29 NAS 1994, pp. 159–160.

30 Albright et al. 1993 provide a country-by-country review of MOX fuel fabrication capacity and plans. See pp. 130–143.

The construction of a MOX plant in Russia would make it more difficult for the U.S. to work toward an interim halt to reprocessing in Russia. Without a MOX plant, reprocessing civilian fuel is only an added expense without any economic return. As in the United States, once the capital investment is made in a large Russian MOX plant, the pressure to use it and to continue reprocessing would be great.

A common goal of stopping reprocessing in Russia could be an important tool for the United States to persuade Britain, France, and Japan to stop reprocessing also. These three countries share the same fears of black markets in plutonium of Russian origin. Fears of similar diversion occurring in Britain, France or Japan are lower at present than for Russia because of the current reality that the first three countries have relatively stable political and economic situations domestically, in contrast to Russia. However, there is no complete guarantee of security of fissile materials in any country.

Dedicated Government Versus Existing Commercial Reactors

The disadvantages arising from potential licensing delays and proliferation concerns associated with commercial reprocessing in other countries could be alleviated to some extent if MOX fuel in the U.S. were loaded into a government-owned reactor whose only purposes were security-related. As noted above, these were the reasons that the OTA study concluded that this would be a more suitable approach than burning plutonium in commercial reactors. Such a reactor would only burn plutonium from the weapons complex and would not be associated in any way with civilian plutonium production or use. A commitment to that effect from the U.S. government could, in theory, reduce the possible concerns of the present or a future Russian government that the U.S. was secretly creating the capacity to use civilian plutonium as a fuel while attempting to set back civilian reprocessing in Russia.

While a government-owned reactor for burning plutonium does resolve some concerns related to licensing and public acceptance within the domestic political framework in the United States, its advantages for non-proliferation are more limited. It also has its own special disadvantages. The first problem is that it would take a long time to convert plutonium into spent fuel, since there would be a very small number of reactors, probably only one, commissioned for this purpose. This means extended storage of plutonium in the United States and hence most likely also in Russia. Second, relations between the United States and Russia are not yet sufficiently stable or friendly for the United States to convince the Russians that the United States did not intend to use its MOX fuel fabrication capacity for civilian plutonium energy production in the long-term. Hence, the United States' ability to dissuade Russia from its present course of plutonium separation would become even more limited than it is today.

There are also special non-proliferation problems associated with a government-owned reactor. It has been proposed from time to time that a reactor dedicated to plutonium burning could also be used to produce tritium for the U.S. nuclear arsenal. Tritium is a radioactive isotope of hydrogen, which is used to boost the yield of nuclear weapons. Since tritium's half-life is 12.3 years, warheads must be replenished with tritium periodically. Due to recent commitments for large nuclear arms reductions, there is a large tritium surplus that can be used for this purpose. A requirement for more tritium for the U.S. arsenal over the next two to three decades would arise only if the United States continues to have a nuclear arsenal of several thousand weapons for decades to come. This would be contrary to the spirit of the NPT and the demands of many non-nuclear weapon states that the nuclear weapon states pursue nuclear disarmament with a definite plan, including a timetable.

The creation of a disarmament plan and timetable will be a difficult enough job. To build a government-owned reactor that could easily be used for tritium production without commitments on further deep arms reductions would increase skepticism as to whether the United States plans to implement its disarmament commitments under Article VI of the NPT.

Foreign Reactors

Several proposals to burn MOX fuel derived from weapons plutonium in foreign reactors have been put forward. They include the use of reactors in Canada, Japan, France, and Germany. These proposals have two basic variants, depending on whether MOX fuel is fabricated in the United States or abroad. MOX fuel fabrication in the United States carries with it the problems that we have already discussed, the most important of which is that it will be very difficult for the United States to convince Russia not to build an infrastructure for civilian plutonium use if it is building one itself. The fact that MOX fuel fabricated in the

United States would be used abroad is not likely to make much difference since once the infrastructure is created, domestic use can follow at little to moderate cost. U.S. MOX fuel fabrication is a part of proposals to burn excess plutonium in Canadian and Japanese reactors.

The option of using foreign MOX facilities has its special security liabilities. The only existing facility large enough to handle the amount of plutonium involved within a relatively short time frame is in Hanau, Germany. To use this facility would be to subsidize plutonium use infrastructure in a country that still has large reprocessing contracts and that is accumulating large surpluses of civilian plutonium. To use the MOX fuel in German reactors would also be problematic since there is no guarantee that the spent fuel produced as a result of MOX fuel use would not be reprocessed. Unless a complete and permanent halt to reprocessing of German spent fuel can be guaranteed and disposition of the large stock of already separated German civilian plutonium (over 14 metric tons by 1990) could be agreed upon, the use of German MOX fuel fabrication facilities poses serious proliferation issues that would be impossible to resolve. Moreover, there is no social or political consensus in Germany that MOX fuel use is wise. In fact, the opening of the Hanau plant has been long delayed due to a protracted federal-state dispute. The use of facilities in France or Japan to fabricate fuel presents similar problems since both countries are strongly committed to reprocessing.

Reprocessing and MOX Fuel Use

The main advantages of MOX fuel with reprocessing cited by its proponents are as follows:

- It creates a fuel source for the long-term.
- It makes use of a fuel resource that has already been created.
- It gets rid of plutonium by fissioning it, which means the high-level waste slated for repository disposal will contain far less long-lived alpha-emitting radionuclides.

As we have already discussed in Chapter 1, plutonium is an economic liability, since the costs of using it are greater than those of using uranium fuel. The potential theoretical advantages of reprocessing are therefore not economically realizable. As a result, expenditures on the use of plutonium either in reactors or any other disposition method has to be justified on security and environmental grounds, since all methods will incur net costs.

Security Concerns

Since reprocessing separates plutonium from spent fuel and puts it back into a weapons-usable form, it will stretch out the security concerns arising from plutonium into the indefinite future. It is a needless expense that exacerbates the geographic spread of plutonium use because plutonium would be used and processed in far more facilities and on a greater scale than it is today.

Environmental and Health Risks of Reprocessing

The environmental costs and risks of reprocessing are also great. There is an extensive literature on this subject, much of which is summarized in a 1992 book, *Plutonium: Deadly Gold of the Nuclear Age*.[31] We will briefly highlight some of the issues here and refer the reader to this book for further details.

Reprocessing involves dissolving spent fuel in acid and then separating plutonium from the other materials by solvent extraction. It takes spent fuel, which is a highly radioactive solid material, and makes it mobile by turning it into a liquid. Reprocessing creates several streams of radioactive materials:

- Plutonium, as a liquid nitrate in the process, and as an oxide finished product;

- Uranium, contaminated with trace quantities of plutonium and some fission products, such as technetium-99;

- Highly radioactive liquid wastes that must be stored in tanks;

- Large volumes of liquid wastes, many of them contaminated to varying degrees with radionuclides, that are released into the environment;

- Releases to the atmosphere of radioactive gases such as carbon-14 in the form of carbon-14 dioxide and krypton-85.

31 IPPNW and IEER 1992.

- Cladding wastes, which derive from stripping fuel from its cladding. The most common kind of cladding wastes consist of radioactive hollow cut-up zirconium alloy pins used in light water reactors.

The total radioactivity of the spent fuel does not depend on whether it is reprocessed. However, the volume of the wastes is greatly increased with reprocessing, and the most dangerous portion of the waste in the short- and medium-term from the environmental standpoint, which consists of highly radioactive fission products, is put into a liquid form which must be stored in tanks. The usual practice in civilian reprocessing has been to store these highly radioactive wastes in acidic liquid form in stainless steel tanks which must be cooled. A loss of cooling due to equipment malfunction or to a loss of electrical power for extended periods could dry out the wastes and cause an explosive release of fission products.

The world's most serious accident in a nuclear weapons plant involved the explosion of a tank containing high-level liquid reprocessing wastes at the Soviet nuclear weapons plant near Chelyabinsk in 1957. The explosion had a force of between 70 and 100 tons of TNT equivalent and caused the evacuation of more than 10,000 people and the long-term contamination of about 6,000 square miles of land.

The liquid waste can be solidified by calcining (roasting) and vitrification. However, reprocessing, liquid waste management, and subsequent vitrification create a whole set of risks in order to create at great expense a fuel that is not only economically worthless but also a threat to security. The final form of plutonium from a civilian reprocessing plant is plutonium dioxide, which is a powder. This can be easily used to fabricate weapons of radiological terror, if diverted.

Finally, the uranium recovered from reprocessing also represents a net liability since it is contaminated with fission products. Processing and enriching this contaminated uranium in chemical plants where natural, unrecycled uranium is processed can greatly increase the costs of final decommissioning of these plants when they are shut down. For instance, technetium-99, which has a half-life of over 200,000 years, often contaminates groundwater, as well as plant equipment.

We will discuss nuclear waste repository related issues in Chapter 6.

MOX Fuel Fabrication

The fabrication of MOX fuel from reactor-grade plutonium presents greater worker health and environmental problems than fabrication of MOX fuel from weapon-grade plutonium. This is mainly because reactor-grade plutonium contains far larger quantities of plutonium-240 and plutonium-241 than weapon-grade plutonium. Plutonium-240 emits neutrons from spontaneous fission. This imposes extra shielding and worker protection requirements. Plutonium-241 decays into americium-241, which is a strong gamma-emitting radionuclide. Americium-241 builds up at a relatively rapid rate due to the short half-life of plutonium-241 (14.4 years).

According to Albright et al., current MOX plants are designed to handle a maximum americium-241 content of 1.3 to 1.5 percent in plutonium separated from spent fuel; this limit is reached for typical light water reactor fuel after two to three years of storage after removal from the reactor. New MOX plants will have a somewhat larger tolerance for up to about 2.5 percent americium-241.[32] This allows an additional two to three years storage time for separated plutonium before fuel fabrication. If the fuel is not fabricated within this time, then the plutonium must be processed to remove the americium-241, making MOX fuel even more expensive and uneconomical.

Conclusion Regarding MOX Fuel

Five countries, France, Russia, Britain, Japan, and India, have civilian reprocessing programs and look upon plutonium as a vital energy resource for the future, even if its economics today look dismal. Another set of countries, Germany, Belgium, Holland, Italy, and Switzerland, own separated plutonium and have contracts with France and/or Britain to provide reprocessing services. Most of these countries are not likely to be amenable to arguments to stop reprocessing or to shelve their plans to acquire more plutonium unless there is clear global leadership on the part of the United States and some guarantee regarding the availability of reactor fuel.

The use of MOX fuel in the United States will necessitate the creation of plutonium handling and MOX fuel fabrication facilities on a large scale. Given that the U.S. long ago gave up civilian reprocessing and that it is highly unlikely to restart it due to the intense political

---

32 Albright et al. 1993, p. 130.

opposition that such a proposal would generate, it would be far freer to use the MOX option domestically if its actions were not so crucial in influencing others. But the United States is the only leading nuclear weapons power that has given up civilian reprocessing for both non-proliferation and economic reasons. It is also the most influential. Therefore, its actions are crucial in shaping international plutonium separation policy. Creating an infrastructure to produce and use MOX in the U.S. will be very detrimental to the goal of stopping reprocessing worldwide. It may be that U.S. influence will not be enough to accomplish this goal, but it will be practically impossible to accomplish without U.S. influence.

The wisest course for the U.S. is to avoid using MOX fuel for plutonium disposition, both in order to more effectively persuade other countries not to do so, and even more importantly, to take more effective global leadership in halting both military and civilian plutonium separation. It should make that decision forthwith.

CHAPTER 4

Plutonium Vitrification: Basic Technical Issues

The principal alternative to converting plutonium to spent fuel in reactors is the vitrification of plutonium, either alone or mixed with other materials. This would put plutonium into a non-weapons-usable form. The degree to which plutonium vitrification will meet the "spent fuel standard" will depend on the materials that plutonium is mixed with and how the processed plutonium is packaged. These factors control to a large extent the resistance to theft and the difficulty of re-extraction of plutonium. In this chapter we explore the basic technical issues related to plutonium vitrification with and without the addition of other materials. In the next chapter we consider the issues related to the use of existing vitrification plants in the United States for plutonium vitrification, the possible construction of new vitrification plants, and sketch the policy implications of these options.

Processing Plutonium as a Waste

The issue of processing plutonium as a radioactive waste can be approached in a manner similar to that for managing high-level radioactive waste. Indeed, spent fuel from nuclear reactors, which contains substantial amounts of plutonium, is classified as high-level waste in the United States. The central difference between plutonium in spent fuel and separated plutonium is that the former is non-weapons-usable material because it needs further complex processing before it can be used in weapons. In contrast, separated plutonium can be used to make weapons and needs to be diluted with other materials before it can be considered a form not usable for weapons without costly and difficult processing. Spent fuel is also very difficult to steal because it emits very high levels of gamma radiation. This difficulty is much reduced if it is stored in shielded casks suitable for transporting it.

Among the different materials that plutonium could in theory be mixed with to render it unusable for weapons are:[33]

- various forms of glass (borosilicate glass, sintered porous high-silica glass, glass ceramic);

- sintered ceramics;

- a synthetic material called SYNROC primarily composed of titanates;

- a number of cement-related materials;

- various metal coatings and other barrier materials.

These same materials, called waste forms or matrices, have been considered for immobilization of liquid high-level radioactive wastes generated by reprocessing, which consist mainly of fission products. Most of the waste forms listed above have not been thoroughly evaluated and manufacturing technologies for many are not fully developed. In 1982, the Department of Energy chose borosilicate glass as the waste form for the waste at Savannah River Plant (now Savannah River Site, or SRS) in large part because the manufacturing technology for glass was far more advanced than that of other proposed waste forms.

Some waste forms other than borosilicate glass potentially possess much better isolation characteristics for reprocessing waste.[34] Several waste forms in which waste oxides are incorporated, such as SYNROC, ceramics or metal matrix waste forms, may possess better isolation properties than borosilicate glass for actinides such as plutonium-239. We have not researched the relative merits of these ceramics versus borosilicate glass as related to the plutonium disposal problem, but it appears that some other glass compositions may be more desirable for reducing long-term releases. Some experimentation has been carried out on glass-ceramics as a waste form at the Idaho National Engineering Laboratory (INEL), where both liquid and calcined high-level wastes are stored. The merits of glass-ceramics for plutonium encapsulation could be explored on an experimental or pilot plant basis. However, borosilicate glass, when

---

33 National Research Council 1983, Table 5.2.
34 National Research Council 1983, Table 5.2.

properly encapsulated and surrounded by engineered barriers, appears to have the potential to contain radioactive waste for very long (perhaps thousands of years or longer) periods in many geologic environments.

Glass has been assessed extensively and possesses the large advantage that it is made using the one well-developed commercial technology that is already available. This has been a major factor in its adoption in the U.S.[35] We will not evaluate the relative cost of glass compared to other waste forms, since the technology for the commercial production of most waste forms is not yet available. As a result, other waste forms would require a long time, perhaps decades, to develop and commercialize; their use would not meet the need to put plutonium into a non-weapons-usable form in as short a time as possible. *This urgency makes the availability of the technology for manufacturing glass the single most decisive factor in the selection of this waste form for plutonium.*[36] Therefore, we have not considered any other waste forms in detail.

Processing Plutonium for Vitrification

The metal form in which plutonium exists in nuclear warheads is too massive, chemically reactive, insoluble in glass, and pyrophoric to be suitable for vitrification. Hence, it must be chemically converted to a more stable form, soluble in molten glass, such as plutonium dioxide.

Plutonium metal can be converted to oxide in one of several ways:

1. Roasting it in an oxidation furnace;

2. Dissolving it in nitric acid in the presence of a small amount of hydrofluoric acid to yield plutonium nitrate, a liquid, which can then be calcined into plutonium oxide;

3. Dissolving it in nitric acid in the presence of a small amount of hydrofluoric acid to yield a plutonium nitrate, which can then be fed into the glass melter, thereby combining the calcining and vitrification steps;

4. Converting it into oxalate (with valence III or IV) and heating it at 1,000° C in air to get plutonium oxide.

---

35 National Research Council 1983, p. 78.

36 There is no comparable urgency for vitrification of high-level waste. The solidification of high-level waste in liquid and sludge forms, which is a matter of some urgency due to the potential for fires or explosions in some tanks, can also be accomplished by calcining or possibly by chemical treatment methods combined with heating.

A new technology, developed at Oak Ridge National Laboratory, allows the oxidation of plutonium to occur after the plutonium and molten glass have been mixed. This is accomplished by adding lead oxide to the melt mix. Plutonium metal reacts with the oxygen in the lead to form plutonium dioxide, while molten lead metal, which is insoluble in glass, sinks to the bottom of the melter. The lead can the then be oxidized again and reused. This method of vitrifying plutonium has been tested on a laboratory scale and may be particularly useful in vitrifying plutonium residues.[37] (See further discussion below.)

Mixing plutonium oxide with glass frit (fragments) in a melter is the final step in vitrification. As noted above, the plutonium may be in oxide or nitrate form at this step. It may be pre-diluted with a neutron absorber, fission products, and/or other actinides. Such mixing may also be combined with the vitrification step (See section on additives below).

A vitrification plant can be designed to cast glass logs, or it can prepare glass frit. The use of glass frit is undesirable since it makes materials accounting extremely difficult, hence complicating the task of safeguarding and verification.

The amount of plutonium per unit weight of glass is an important parameter in vitrification because it determines the number of glass logs that have to be cast. It therefore affects the cost and length of time for vitrifying a specified quantity of plutonium. A high concentration of plutonium per unit of glass would enable faster completion of vitrification. However, a high concentration reduces the cost of recovery, and hence has disadvantages from the proliferation and reuse points of view.

The limits of plutonium loading per unit weight of glass depend on a number of considerations. The primary ones are:

- solubility of plutonium in glass;
- criticality;
- recoverability of plutonium once the glass has been made.

We will discuss the last two issues below in the section on additives. First let us briefly consider the question of solubility limits.

The solubility of plutonium depends on the specific composition of the glass. Berkhout et al. have cited a German experiment that found a solubility limit of 4.5 percent of plutonium in glass.[38] Work at the

---

[37] Forsberg et al. 1994.
[38] Berkhout et al. 1992, p. 30.

Savannah River Laboratory indicates a solubility limit in borosilicate glass of 6 to 7 percent for plutonium in the IV valence state, which would apply to plutonium dioxide. The limit may be even higher for valence states III, V, or VI.[39]

Figure 1 shows a flow diagram of the steps for alternative routes to plutonium vitrification. The steps for adding various materials to the glass so as to achieve desirable glass properties are not shown in this figure.

Additives in Plutonium Vitrification

In contrast to vitrification of high-level waste, it may be desirable to mix plutonium with other materials prior to or during vitrification. There are several objectives for such mixing:

1. Reducing criticality concerns by adding boron and/or another neutron absorber;

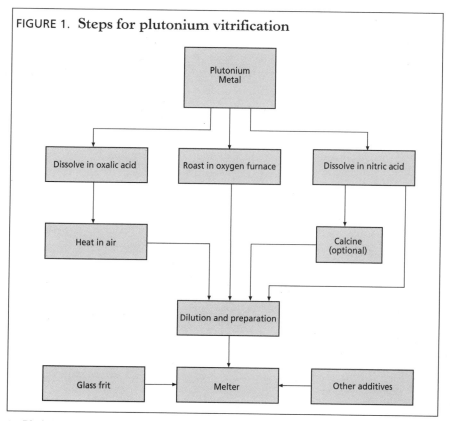

FIGURE 1. Steps for plutonium vitrification

[39] Plodinec 1979.

2. Making it more difficult and costly to extract plutonium from the waste form by mixing in fission products and/or other materials;

3. Improving glass characteristics.

We will briefly consider each of these reasons for adding materials to the plutonium vitrification process.

Criticality Concerns

A criticality occurs when plutonium (or other fissile material such as uranium-235) achieves a self-sustaining chain reaction. In this reaction, at least one neutron from the fission of a plutonium nucleus splits one other plutonium nucleus, on average. The reaction therefore continues once started by a source of neutrons so long as the criticality persists. Some free neutrons are always present in plutonium due to spontaneous fission of a small fraction of its nuclei.

A criticality can release a large amount of energy due to a large number of nuclear fission events occurring in a short period of time. However, there is no practical concern that the plutonium would explode like a nuclear weapon. This is because once a criticality is initiated it would cause the critical mass to fly apart after a small to moderate amount of energy has been released. However, the potential level of energy release can be high enough to cause threats to worker safety and to the structural integrity of the system of plutonium containment.

A minimum of 2 percent plutonium is necessary for a critical mass in a light water reactor core.[40] However, if plutonium is added to borosilicate glass, the percentage of plutonium in glass can be increased considerably without creating the risk of criticality, since boron is a neutron absorber. According to a report of the Plutonium Vitrification Task Group of the Westinghouse Savannah River Company, 7 percent of plutonium by weight in borosilicate glass appears to be criticality safe.[41]

Seven percent by weight is the upper limit of plutonium solubility in borosilicate glass. But if the concentration of plutonium in glass is as high as seven percent, it may be too easy to recover it and hence make it weapons-usable again. As a result, the limit for plutonium concentration in borosilicate glass is not determined primarily by short-term criticality

---

40 Berkhout et al. 1992, p. 31.
41 Westinghouse 1993, p.2.

considerations.[42] However, it is still necessary to take proper precautions to avoid accidental criticality when preparing and feeding the plutonium into the glass melt. One way to address these concerns would be to pre-mix the plutonium and boron (and/or another neutron absorbing material—see below) prior to feeding the mixture into the melter.

Glasses other than borosilicate glass may require small amounts of boron and/or other neutron absorbers to be added to avoid criticality risks, depending on the concentration of plutonium in the glass.

An additional criticality concern arises because plutonium-239 decays into another fissile material, uranium-235. This is an environmental and licensing concern for repository disposal. We will discuss this in Chapter 6.

Increasing the Difficulty of Plutonium Recovery

If plutonium is vitrified by itself, its susceptibility to being separated from the glass becomes a concern. For instance, the glass could be crushed and a heated nitric acid solution (90 to 95°C) added to it. Laboratory experiments have demonstrated that more than 99 percent of the plutonium can be dissolved in this fashion, along with most of the glass components. In the resulting solution a gelatin "cake" would form containing the silicon from the dissolved glass. The silica would then be removed using a centrifuge process that separates lighter from heavier molecules. Subsequent steps would be similar to chemical reprocessing, in which plutonium is purified by separating it from various other elements.[43]

Mixing plutonium with actinides

In order to make recovery of vitrified plutonium difficult and costly, plutonium may be mixed with other materials prior to or during vitrification. For example, vitrifying plutonium mixed with heavy elements such as uranium would make plutonium more difficult to recover, since it would add steps to the process of separating the plutonium from the mixture of elements. Uranium belongs to a series of elements called *actinides* which have similar chemical properties. Plutonium is also an actinide. Therefore, separating plutonium from uranium is chemically a difficult task. Thorium-232 is also an actinide and could potentially be used as a diluent.

---

42 See Chapter 6 for a discussion of long-term, repository-related criticality concerns.
43 Westinghouse 1993, p.7.

Mixing depleted uranium in a ratio of ten or more parts of uranium to one part of plutonium would render the recovered uranium-plutonium mixture practically impossible to use for nuclear weapons. This is because depleted uranium consists almost entirely of uranium-238, which cannot sustain a chain reaction. The separation of uranium and plutonium in stages similar to those used in reprocessing spent fuel would then be required for producing weapons-usable material. The main difference between reprocessing spent fuel and a vitrified uranium-plutonium mix is that, unlike glass with uranium and plutonium in it, spent fuel emits very strong gamma radiation, and hence the reprocessing of spent fuel must be done remotely. Depleted uranium mainly emits alpha radiation, which cannot penetrate the dead layers of skin, and so is not as dangerous outside the body as intense gamma-emitting radionuclides; but uranium can be very damaging once it is inside the body (via inhalation, for instance).

If substantial quantities of heavy elements such as depleted uranium are added to borosilicate glass, it may be necessary to adjust glass composition so as to minimize the potential for radionuclide releases, especially if there is evidence that these materials may be an important component of potential radiation doses to the public. Depleted uranium may be a good diluent in this regard, since it has a very low specific activity (radioactivity per unit weight) compared to many other actinides.

The use of thorium-232 may be more advantageous in that thorium is a closer chemical analog of plutonium than is uranium. However, the decay products of thorium-232 build up faster than those of uranium. As a result, thorium-232 processing is generally more costly and difficult than equivalent processing of depleted uranium. This could, of course, also be seen as a non-proliferation advantage. The relative merits of thorium-232 versus depleted uranium need to be carefully examined, with due consideration to safety, speed, and non-proliferation concerns.

Mixing plutonium with non-radioactive elements

Depleted uranium is not the only material that might be considered for mixing with plutonium in order to render the mixture extracted by dissolution in acid unusable for weapons. Non-radioactive elements might also be considered. In particular, non-radioactive isotopes of a class of elements known as "rare earths" could be attractive because they have chemical properties similar to elements in the actinide series, including plutonium. The use of these elements has been recently investigated by Kory Sylvester at the Massachusetts Institute of Technology under grants from the DOE and the W. Alton Jones Foundation.[44]

The research, which included computer modeling and laboratory work, indicates that mixing plutonium with rare earths, particularly europium, would increase the difficulty of re-extraction of plutonium, especially for sub-national groups. Even if the plutonium-europium mixture was separated from the glass by chemical processing, the critical mass of the mixture would be very large, so that building a nuclear weapon would become impractical or impossible without further expensive processing.[45]

The advantages of mixing plutonium with a rare earth element are similar to those for mixing it with depleted uranium. The difficulty of re-extraction would be greatly increased for sub-national groups, but governments that now have separated plutonium would be able to recover it somewhat more easily than plutonium vitrified with high-level wastes.

It should be noted that the addition of depleted uranium, europium or other diluents would require a larger amount of glass than the vitrification of plutonium alone, not only due to the solubility limits of these materials but also because of the need for large amounts of diluent for non-proliferation goals. This means that larger capacity for glass production must be installed compared to vitrification of plutonium alone or with fission products. Despite this, vitrification with an actinide or with a non-radioactive rare earth element could be accomplished more rapidly than vitrifying plutonium with fission products in a new plant.

Mixing plutonium with fission products

Mixing plutonium with a sufficient quantity of one or more gamma-emitting fission products makes recovery of plutonium comparable in difficulty to reprocessing spent fuel. This is because considerably greater radiological protection would be required for handling and recovering the plutonium, more recovery steps and equipment would be needed, and provisions would have to be made for storing highly radioactive waste containing the fission products that would remain after the plutonium has been re-extracted. Further, since the glass logs would be highly

---

44 Sylvester 1994.

45 Sylvester 1994, Chapter 4.

radioactive, they would be very difficult to steal, giving such vitrified plutonium essentially all the advantages of the spent fuel standard.[46]

By the same token, it is also more costly and complicated to vitrify plutonium mixed with highly radioactive wastes or even a single gamma-emitting radionuclide like cesium-137. The shielding requirements and concomitant need for remote ("hot cell") operation give rise to far greater construction and operating costs. This also means a greater lead time in designing and constructing such a facility. It would likely take more than a decade to complete the environmental reviews and build a suitable plant, even if no pilot plant were built.

Cesium-137 may be a suitable fission product to combine with plutonium for vitrification because it is an intensely radioactive gamma emitter with a moderate half-life (about 30 years). Consequently, using cesium-137 in the plutonium vitrification mix would make recovery of the plutonium from the glass very difficult for a few hundred years. However, since the half-life of plutonium-239, at 24,110 years, is so much longer than cesium-137, it would be possible after several hundred years to recover essentially all the plutonium from glass with the same level of difficulty as if the glass had not been mixed with cesium-137. In other words, mixing plutonium with a fission product like cesium-137 is only different from vitrifying plutonium alone for a few hundred years. Cesium-137 for mixing with plutonium is available as separated material stored in capsule form at DOE's Hanford, Washington facility. The calcined mix of fission products stored at Idaho National Engineering Laboratory could also be used.

The fact that adding cesium-137 would make plutonium extraction very difficult and complex for a few hundred years could allow sufficient time to develop long-term disposal methods and would reduce the risk of the plutonium being recovered in the interim.

Since the difficulty of recovering plutonium mixed with fission products and vitrified in borosilicate glass is comparable to recovering it from spent reactor fuel, it conforms to what has come to be called the "spent-fuel standard" for treating plutonium as a waste. After disposal in a repository, the costs of recovering the glass canisters from a sealed

---

46 There is some debate whether plutonium vitrified with fission products is truly equivalent to spent fuel. This is because weapon-grade plutonium remains after vitrification, whereas reactor-grade plutonium results from running weapons-plutonium through a reactor for a sufficient length of time. Since it is more difficult from the point of view of handling and safety to process reactor-grade plutonium, there is some justification in this claim. It does not appear to us that this makes a crucial difference from a non-proliferation perspective, which is the central thrust of the spent fuel standard, since reactor-grade plutonium can be used to make nuclear weapons.

repository and re-extracting the plutonium would probably be far larger than new plutonium production.

The more dilute the plutonium in the glass and the greater the quantity of contaminants added, the more difficult and costly recovery of plutonium becomes. However, the amount of glass to be manufactured is inversely proportional to the plutonium concentration: the lower the concentration of plutonium, the more canisters and repository space are needed for disposal, with the consequent likelihood of longer vitrification time and cost increases.

An advantage of adding fission products to plutonium that may offset the higher costs to some extent is that it would allow for concentrations of plutonium in glass of several percent and still conform to the spent-fuel standard. Such high concentrations of plutonium would mean fewer glass canisters would have to be produced to immobilize a given amount of plutonium. This would reduce the time needed to vitrify plutonium and the space needed for disposing of the glass canisters in a repository.

A combined option

Each of the options we have discussed possess significant advantages and disadvantages. The advantages of the various options can be combined if plutonium is vitrified with an actinide or a rare earth and *the canister* is made radioactive with a gamma-emitting fission product such as cesium-137. This would provide all the advantages of rapidity because the vitrification plant would not handle radionuclides that emit strong external gamma radiation. Therefore, the plants could be built in a shorter time and not require hot cell construction. The cesium-137 could be added to the canister prior to sealing it in a special facility. Alternatively, a second outer canister alloyed with cesium-137 could be fabricated. About 135 grams of cesium-137 (about 12,000 curies) would be required to produce the same external gamma radiation field of 5,500 rads as would be produced by roughly 200,000 curies of Savannah River high-level waste mixed with glass and cast into a canister typical of that planned for the Savannah River Site vitrification plant (see Chapter 5.)[47] The operation of adding the cesium-137 could be done well after the vitrification.

Adding a gamma emitter to the canister would provide essentially the same resistance to theft as the vitrification of plutonium with the fission

---

[47] John Plodinec, personal telephone communication to Noah Sachs, November 28, 1994.

products. But a far smaller quantity of the fission product is required to produce the same external radiation field because it is present in the outer layer only rather than throughout the glass volume. This would reduce worker exposure for a given level of resistance to theft. While the proliferation resistance would approach that of vitrification with fission products in almost all respects, the plutonium re-extraction *costs* would be lower for several reasons. For instance, the quantity of high-level radioactive waste produced would be far smaller. Lower costs make this option more compatible with providing a financial guarantee for re-extraction of plutonium for reuse as a fuel relative to vitrification with fission products.

Table 3 shows a comparison of vitrification options with respect to each other (and *not* relative to any other option discussed in this report):

TABLE 3. Comparison of vitrification options

OPTION	ADVANTAGES	DISADVANTAGES
1. Vitrification of plutonium alone	Simplest and most rapid option	Least technical difficulty for plutonium re-extraction; low resistance to theft
2. Vitrification of plutonium with fission products	Highest initial proliferation-resistance both as regards difficulty of theft and of re-extraction	May hamper global agreement on an interim halt to reprocessing; likely to take the longest time; in a few centuries proliferation resistance declines to approximately that of vitrification of plutonium alone.
3. Vitrification of plutonium with actinides or rare earths	Moderate to high technical proliferation resistance; can be done rapidly; durable proliferation resistance	Low resistance to theft.
4. Option 3 with a gamma emitting canister	High technical proliferation resistance; can be done rapidly; durable proliferation resistance; high resistance to theft	Re-extraction less difficult than with Option 2. This could be an advantage. See text.

Vitrification of Scrap and Residues

A large amount of plutonium in the form of scrap and residues with high plutonium content is created during plutonium processing for nuclear weapons production. When new weapons were being produced, scrap residues with a high plutonium content were processed and the plutonium was recovered for use. In some ways, the plutonium in scrap and residues is now even more of a security, environmental, and economic liability than metal plutonium pits from dismantled warheads. This is because such plutonium is far less well accounted for, often stored in deteriorating containers, and sometimes mixed with flammable substances. Some residues are in liquid form, such as plutonium nitrate, posing serious environmental and safety threats.

The amount of plutonium in the U.S. nuclear weapons complex that has been transferred from the Defense Programs division to the Environmental Management division is about 33.4 metric tons. Of this amount, 16.3 metric tons are in the form of plutonium metal or oxide. About 9.4 metric tons consists of scrap, residues, and plutonium solutions; 7.4 metric tons are in unreprocessed spent fuel, and the rest, amounting to about 0.3 metric tons, is in other forms and sealed sources. In general, the scrap, residues, solutions, and deteriorating spent fuel rods pose the most serious near-term safety problems.[48] But even some of the plutonium stored in metal form presents severe hazards because it was improperly packaged in plastic sheets prior to being inserted in metal storage cans.[49]

Stabilizing these forms of plutonium poses one of more urgent safety and health problems in the nuclear weapons complex. There is currently no operational plan to reduce the hazards of storage in the interim for these unstable forms of plutonium, though the DOE has tried to develop an "interim storage guidance." The Defense Nuclear Facilities Safety Board, the official watchdog of the DOE in the absence of external regulation, has noted that:

> There is a problem with the interim [storage] guidance. It does not acknowledge the reality that much of the current plutonium inventory is in forms that are unsuitable for even interim storage. The interim guidance simply states that solutions should not be stored, even though three of the four main sites have solutions in storage. It

---

48 Grumbly 1994.

49 *New York Times*, December 7, 1994.

does not mention reactive scrap at all. DOE could provide guidelines on how best to monitor reactive scrap materials until they can be processed and eliminated. DOE can also provide guidelines on how to decide whether to repackage reactive scrap.[50]

To address these issues, DOE is conducting a plutonium "vulnerability assessment" due to be completed in January 1995, as this book goes to press. The assessment was undertaken because ruptures were discovered in stored plutonium packages.[51] A draft made public in December 1994 "identified 299 environment, safety and health vulnerabilities at 13 sites" with the most important locations being the Rocky Flats Plant near Denver, Colorado, the Hanford Site in eastern Washington state, the Savannah River Site in South Carolina, and the Los Alamos National Laboratory in New Mexico.[52] The report acknowledges the seriousness of the problems associated with storage of certain forms of plutonium and the urgent need for safer storage:

> Plutonium package failures and facility degradation will increase in the future unless problems are addressed in an aggressive manner. The Department needs a strong, centrally coordinated program to achieve safe interim storage of plutonium. Priority must be given to plutonium solutions, chemically reactive scrap/residues and packaging with plastics or other organics.[53]

Due to the serious health risks involved in processing plutonium-bearing materials, the DOE should consider how it might stabilize these materials and make them suitable for extended storage in as few steps as possible. From this point of view also, a considerable investment to complete the development of the technology developed by Oak Ridge or some other approach similar to it is warranted (see below).

We have no estimates for the corresponding figure for scrap and residues in Russia, but it is likely to be as large or larger and even less well accounted for. The accounting of this material, its processing into safer forms when necessary, and safer storage are critical for improving the environmental and security outlook as regards weapons-usable fissile materials.

The direct vitrification technology developed at Oak Ridge National Laboratory mentioned above has the potential for vitrifying a large

50 DNFSB 1994, p. 11.
51 DOE 1994a, p. vi.
52 DOE 1994a, p. vi.
53 DOE 1994a, p. vii.

fraction of plutonium scrap and residues if the technology can be successfully scaled up to a pilot plant level and demonstrated. The U.S. should also consider sharing this technology with Russia as part of a larger understanding to account for and put plutonium now in scrap and residues into verifiable storage. One approach for doing this and for building pilot plants with public participation is discussed in Chapter 8 on policy issues.

Environmental Controls and Worker Health and Safety

Environmental controls and worker protection will be needed for all disposition options, because plutonium is highly carcinogenic. In the case of vitrifying plutonium mixed with fission products, the measures needed would not be greatly different from those required for the vitrification of the wastes without the addition of plutonium. The processing of plutonium to prepare it for vitrification will require safety measures comparable to those at fabrication plants for the plutonium components in nuclear weapons in order to guard against possible internal exposures.

Adding gamma emitters such as cesium-137 makes it far more complex and costly to vitrify plutonium than to vitrify plutonium alone. This is due to the need for massive shielding, remote handling and extra monitoring equipment that accompanies remote handling. The level of shielding and size of the plant are far smaller when plutonium is vitrified alone, with depleted uranium (or similar materials), or non-radioactive chemical elements. This means that plants can be built faster. Small-scale, modular plants are also more feasible with these approaches. Such plants may be advantageous in the processing and disposition of plutonium residues. These residues include plutonium-bearing materials in various chemical and physical forms, such as materials held-up within processing equipment that were not fully processed into final forms. Such residues must be dealt with as part of the decommissioning of several weapons plants.

The vitrification of separated reactor-grade plutonium with uranium or rare earths is likely to require more worker protection measures and shielding than the same operations for weapon-grade plutonium. This is because the americium-241 content of aged separated reactor-grade plutonium makes it a far stronger gamma-emitting material. Additional protection is also require from neutrons originating in the spontaneous fission of plutonium-240.

Adding cesium-137 or equivalent fission products, such as calcined radionuclides from the Idaho National Engineering Laboratory, to the canister would reduce worker exposures for a given external radiation field because the amount of fission products to be handled would be much reduced. There would likely be increased transportation requirements under this option because calcined fission products and plutonium are stored at different locations.

Accidents

Apart from criticality accidents mentioned above, other types of accidents, such as spills of radioactive materials, might occur during the vitrification of plutonium. It is expected that these would be similar to those that any vitrification operation for high-level wastes might experience. Such accidents have been studied in the context of the construction of the vitrification plants at Savannah River Site and at West Valley (see Chapter 5). There may be some additional risks and costs arising from the addition of alpha-emitting plutonium. Plutonium vitrification options need to be studied in an Environmental Impact Statement under the National Environmental Policy Act so that a sound decision regarding the choice(s) of vitrification technology can be made. More than one vitrification technology may be needed since there are many different forms of plutonium to be dealt with.

CHAPTER 5

Facilities for Vitrification and Plutonium Processing

Existing Vitrification Plants

There are two vitrification plants in the United States: one at the Savannah River Site, called the Defense Waste Processing Facility (DWPF), and the other at West Valley, New York, called the West Valley Demonstration Plant (WVDP). The former was built for vitrifying high-level waste from military plutonium and tritium production; the latter was built to process high-level waste mainly from commercial nuclear reactor spent fuel, though some of the high level waste in West Valley is also of military origin.

DWPF - Savannah River Site

The DWPF at the Savannah River Site is a plant built to vitrify a portion of the 34 million gallons of high-level wastes stored in 50 tanks on site.[54] The wastes are to be pre-treated and concentrated prior to vitrification. The wastes resulting from pre-treatment, which will contain some long-lived radionuclides such as technetium-99, will be mixed with cement and poured into low-level waste disposal facilities called "saltstone vaults" on site. Figure 2 shows the flowsheet for the high-level waste solidification program at the Savannah River Site.

The feed to the melter will contain only 40 percent solids; in effect, the melter in which the wastes are mixed with molten glass will also act as an evaporator for the liquids, which consist mainly of water. At 85 percent fill, each canister will contain 625 liters of glass. The diameter of the canister itself will be about 61 cm, and its height about 3 meters.

---

54 There are 51 tanks, but one was emptied after it leaked and contaminated the surrounding soil.

The empty weight of each canister will be about 500 kilograms and each will contain 1,682 kilograms of glass.[55] The density of the glass is about 2.7 grams/cc. The capacity of the melter is about 100 kilograms of glass per hour. Figure 3 shows a diagram of the DWPF melter and Figure 4 shows the specifications of a DWPF glass-containing canister.

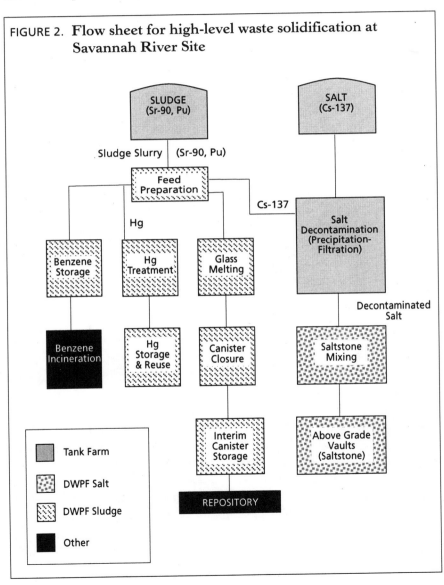

FIGURE 2. Flow sheet for high-level waste solidification at Savannah River Site

---

55 ORNL 1992, vol. 1, Table 3.1.1.

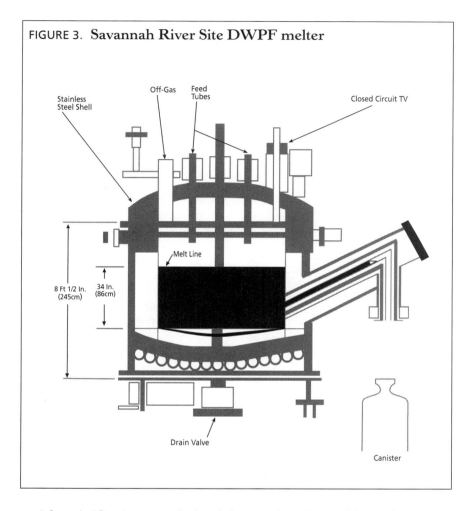

After vitrification, a typical stainless steel canister will contain a maximum of 234,000 curies of radioactivity at the time of fill, primarily in fission products, of which strontium-90 and cesium-137 make up most of the radioactivity.[56] There is likely to be a considerable variation in the amount of radioactivity and specific fission products from one canister to the next due to the non-uniformity of the feed material. According to J.M. McKibben of Westinghouse Savannah River Company, the high-level waste at Savannah River will be vitrified in 6,100 canisters,

---

56 High-level wastes when first discharged from a reprocessing plant typically contain large quantities of relatively short-lived fission products such as ruthenium-106. These will have almost entirely decayed away by the time the waste at Savannah River Site is vitrified.

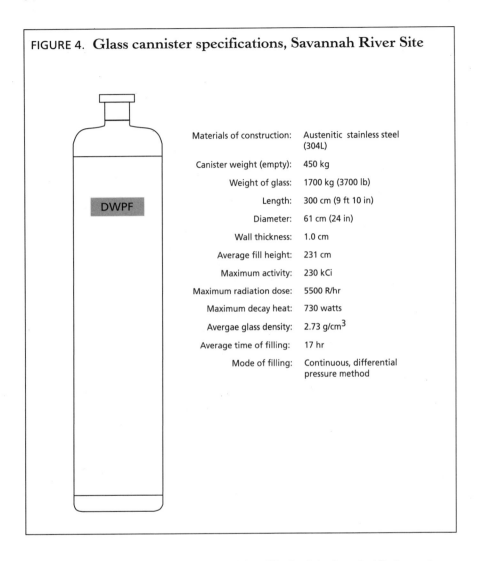

FIGURE 4. **Glass cannister specifications, Savannah River Site**

Materials of construction:	Austenitic stainless steel (304L)
Canister weight (empty):	450 kg
Weight of glass:	1700 kg (3700 lb)
Length:	300 cm (9 ft 10 in)
Diameter:	61 cm (24 in)
Wall thickness:	1.0 cm
Average fill height:	231 cm
Maximum activity:	230 kCi
Maximum radiation dose:	5500 R/hr
Maximum decay heat:	730 watts
Avergae glass density:	2.73 g/cm^3
Average time of filling:	17 hr
Mode of filling:	Continuous, differential pressure method

perhaps a "few hundred" more will be filled with the vitrified products resulting from cleaning out and decommissioning the vitrification plant itself.[57]

Savannah River Site tanks contain substantial quantities of both plutonium-238 (about 1.5 million curies) and plutonium-239 (about 27,000 curies). The weight of these isotopes is about 86 kilograms and 430 kilograms respectively, yielding a total of about 516 kilograms of plutonium.

57 J.M. McKibben to Peter Johnson, personal communication, April 30, 1993, p. 3.

Facilities for Vitrification and Plutonium Processing

As noted above, about 6,100 canisters will be cast, each containing about 1,682 kilograms of glass. Thus, there would be an average of about 0.09 kilograms of plutonium in each canister produced from existing high-level waste, amounting to only 0.005 percent of the weight of the glass. This is far lower than the concentrations of 1 to 5 percent being discussed for disposal of plutonium from dismantled warheads. Therefore the variation in the amount of plutonium from one canister to the next due to the varying concentration of plutonium in different tanks will not add significantly to the number of logs that must be cast to keep weapons-plutonium in the glass at a particular concentration.

The maximum heat output per DWPF canister is expected to be about 730 watts. Adding one percent weapon-grade plutonium to high-level waste would increase decay heat by about 37 watts per canister, or about 5 percent.[58] At this level of plutonium concentration the heat load would not be significantly increased. However, addition of several percent weapon-grade plutonium would increase the heat load considerably, notably in canisters that have a heat loading well below the maximum projected level. In the long-term (more than a few hundred years), plutonium would come to be the dominant heat source since cesium-137 and strontium-90 each decay with a half-life about 30 years; they will be essentially dissipated after about ten half-lives.

There is, therefore, in theory, a considerable amount of flexibility to add a substantial amount of plutonium per canister before hitting solubility or heat loading limits. One percent plutonium by weight would mean the addition of about 17 kilograms of plutonium (about 19 kilograms of plutonium dioxide) to each canister. At that rate, it would take up to about 3,000 canisters to accommodate up to 50,000 kilograms of plutonium.

In principle, one could add five percent, or about 85 kilograms, of plutonium to each canister before approaching plutonium dioxide solubility limits. Therefore, it is theoretically possible to accommodate 50,000 kilograms of plutonium in about 600 canisters.

The initial annual projected production rate of DWPF is 410 canisters per year, so 50 metric tons of plutonium could be vitrified in periods ranging from just under one-and-a-half years (in the case of 600 canisters) to a little over 7 years (in the case of 3,000 canisters), depending on the plutonium concentration in the glass. This production rate

---

58 Berkhout et al. 1992, Table 5 and p. 31. Isotopes with shorter half-lives, such as plutonium-241, produce more decay heat because they decay faster.

could be increased considerably if dry rather than wet materials are fed into the melter. (The latter is currently planned for high-level waste.)

WVDP-West Valley

There are 2,270 cubic meters of alkaline high-level waste and 45 cubic meters of acidic high-level waste at West Valley.[59] Most of the volume is considered "low-level" radioactive waste; it is being mixed with cement and formed into 71-gallon square drums which are stored above ground. The bulk of the long-lived radioactivity (apart from technetium-99) is to be incorporated into 300 glass canisters of approximately the same dimensions as the DWPF canisters. The WVDP melter has about half the capacity of the DWPF melter.[60]

Cement drums are being produced at West Valley and, according to present plans, vitrification will commence in 1996 after cement drum production is finished.

The total radioactivity in the high-level waste at West Valley is about 27 million curies, which means that there will be about 90,000 curies per canister,[61] with a heat output of about 300 watts. The present plan is to retire the WVDP melter after the 300 canisters are produced. Presuming that the annual production capacity is about 200 canisters per year (half of DWPF), only a portion of a consignment of 50 metric tons of plutonium could be vitrified at West Valley. If the plutonium concentration is low, on the order of 1 percent or less, then only a small fraction of 50 metric tons could be vitrified at West Valley. At 5 percent concentration about 25 metric tons of plutonium could be vitrified in the planned production run. However, in theory, the production at West Valley could be extended, and cesium-137 from the capsules stored at Hanford could be mixed with the plutonium to provide a waste form with sufficiently high gamma radioactivity to meet the spent fuel standard.

Table 4 summarizes the range of schedules for vitrification of 50 metric tons of plutonium under various assumptions about plutonium content per canister. We assume that the production rate at DWPF is 400 canisters per year and that at WVDP it is 200 canisters per year.

59 ORNL 1992, vol. 3, p. 3B-3.
60 WVDP 1990.
61 OTA 1991, Figure 1-6.

TABLE 4. **Time in years to complete vitrification of 50 metric tons of plutonium**

AMOUNT OF Pu	DWPF Pu=1%	DWPF Pu=5%	WVDP Pu=1%	WVDP Pu=5%
50 metric tons	7.5	1.5	15	3.0

Note: We assume the following for output: DWPF = 400 canisters per year; WVDP = 200 canisters per year.

Table 4 shows that vitrification of 50 metric tons of plutonium can be accomplished in under one decade, and, if the plutonium concentration is sufficiently high (5 percent), in less than five years, even with a single melter.

Some Practical Considerations Regarding the Use of DWPF and WVDP

DWPF

DWPF is seriously behind schedule for several reasons, including a number of technical difficulties associated with preprocessing high-level waste prior to vitrification. It is currently scheduled to start up in December 1995. Therefore there is a theoretical possibility that vitrification of plutonium could begin soon thereafter.

Some problems may prevent this option from being realized in practice. In October 1992, M.J. Plodinec, Manager of Glass Technology of the Savannah River Glass Technology Center, noted that the "additional infrastructure (e.g., critically safe feed pretreatment), and paperwork (e.g., major revision of the SAR [Safety Analysis Report]), and necessary pit processing" would cause additional delays in plant start-up.[62]

Additional modifications may be needed for environmental reasons. The DWPF is equipped with an elaborate system of environmental controls, including HEPA (High Efficiency Particulate Air) filters. Presuming that this system functions as designed and predicted, the processing of plutonium may not add significantly to expected impacts from high-level waste vitrification. However, according to J.M. McKibben of

[62] M.J. Plodinec, personal communication, October 12, 1992.

Westinghouse Savannah River Company, the "volatility of [plutonium] in the melter . . . would require DWPF design modifications, including adding neutron monitors at several locations to detect [plutonium] accumulations."[63]

One serious disadvantage of adding plutonium vitrification to the DWPF function is the delay in the vitrification of the Savannah River Site high-level wastes that would likely result. A large portion of these wastes are stored in liquid and sludge forms, which are highly undesirable from environmental and safety standpoints, since there is far greater potential for environmental contamination from their storage (for instance, from tank leaks, fires, or explosions) than from storage of vitrified wastes. The consequences of such accidents with liquid wastes would also be far more severe than those of accidents that might occur from storage of vitrified high-level waste.

Further, the DWPF has faced repeated and serious delays in part because no pilot plant was built prior to building a full scale plant for processing and vitrifying Savannah River Site high-level wastes. Indeed, DOE and Du Pont, its contractor for the plant until the late 1980s, thought fit to proceed with the design and construction of a full-scale plant without ever casting a single full-size radioactive waste canister with real radioactive waste.

Adding facilities to handle plutonium and to satisfy associated criticality and safeguards requirements would add another layer of complexity to an already complex and troubled plant. It is unclear that the plant could in practice survive either the engineering challenges or the environmental and safety reviews that would accompany the addition of plutonium vitrification. In view of the many problems that DWPF has already faced, a very crucial security issue, the vitrification of plutonium, should be added to its function only after very careful consideration of DWPF's drawbacks. Plutonium disposition should not be allowed to compromise the already large and urgent mission of vitrifying high-level waste that the plant must carry out successfully. Spending many years trying to modify a plant with a history of delays and unanticipated problems could jeopardize two centrally important public policy goals from being accomplished, instead of just one, should the plant not perform as planned.

DWPF could also vitrify plutonium alone or in combination with

---

63 J.M. McKibben to Peter Johnson, personal communication, April 30, 1993, p. 3.

Facilities for Vitrification and Plutonium Processing

depleted uranium or other heavy metals discussed in Chapter 4. According to Westinghouse, such processing could begin "as early as 2002 with additions and modifications of these facilities" in the F-area as well as changes to DWPF.[64] The F-area is one of the two areas at Savannah River Site where reprocessing and plutonium processing facilities are located. (The other area is called the H-area.) However, the problem of vitrifying high-level waste, which is also an urgent one, may consequently receive a setback. Since vitrification of plutonium alone or mixed with heavy metals that are not strong gamma emitters could be accomplished in newer melters more rapidly than in DWPF (see below), use of DWPF for this purpose does not appear advantageous, especially since it could adversely affect the important goal of high-level waste vitrification.

One possible route to using DWPF could be to plan to start it up for its original purpose of vitrifying high-level waste in December 1995 and to begin careful feasibility studies on paper and in the laboratory on the addition of plutonium to the mission of DWPF. Since the present melter must be replaced in a few years in any case, the plutonium vitrification function could be incorporated into the design of the new melter and added at that time. Another possibility under consideration is the construction of facilities adjacent to the DWPF for plutonium vitrification.

In sum, while the existing DWPF melter should not be used for plutonium vitrification, there might be a number of ways in which existing facilities might be used to carry out some portion of the plutonium vitrification function at the Savannah River Site.

WVDP

The use of WVDP for plutonium vitrification faces similar obstacles, but has some additional ones as well. J.M. McKibben notes that there are many obstacles to the vitrification of plutonium at West Valley:

- The site is owned by the state of New York and not the federal government;
- The present vitrification plant has been designed for a short life; extending its operation might result in environmental and health risks;

---

64 Westinghouse 1993; p. 2.

- It is neither licensed nor designed to handle criticality risks associated with plutonium;
- It does not have the infrastructure and support capability to comply with DOE requirements for Special Nuclear Material Safeguards.

In sum, there are many serious practical drawbacks to using existing vitrification plants for vitrifying plutonium. Even so, the use of DWPF for vitrifying plutonium in ways that do not delay high-level waste vitrification should be explored. The NAS study concluded that, while some technical issues remained to be resolved, the use of DWPF was promising enough, at a moderate enough cost (about $1 billion for 50 metric tons of plutonium) that it should be considered as a leading option. (The other option, according to the NAS study, is the use of plutonium as MOX fuel in existing reactors.)

New Vitrification Plants

We will now explore the option of building new modular plants. The modular approach has several advantages. Among the chief ones is that failure of one melter or an associated system component does not result in a shutdown of the entire process of plutonium vitrification. One or more new vitrification plants could be built either to vitrify plutonium alone or to vitrify plutonium mixed with radioactive materials. We will consider both these options.

Vitrification Mixed with Dry High-Level Waste

If a new plant to vitrify plutonium with highly radioactive waste is built, it would be best to avoid using existing liquid, sludge, or salt wastes in tanks, for reasons we have already discussed. Such a plant should be based on mixing plutonium with separated dry cesium-137 from Hanford, or with dry, calcined wastes now stored at INEL. In view of the practical difficulties faced by DWPF, the most immediate technical decision involving plutonium vitrification is whether to build a pilot plant. This is a complex political and technical issue. Plants to vitrify high-level wastes in borosilicate glass are operating in other countries, notably France and Britain. Unlike the single-step drying and vitrification design of DWPF, the French design involves two steps. First the waste is calcined, and then the dry waste is vitrified. Therefore there is already a considerable amount of practical experience in vitrifying calcined wastes of the type that would be mixed with plutonium in a new plant. However, as a result of past policy mistakes, precious little of that experience exists in the United States.

Even with the incorporation of foreign experience with high-level waste vitrification, there still is the question of integrating the handling of large amounts of plutonium with vitrification technology. This introduces a sufficient number of new issues, such as criticality and safeguards concerns, so that we feel a pilot plant is, in any case, necessary. As a result, added to the necessary time it would take to build a new full-scale plant, including completing required environmental reviews, it would probably take five years (and perhaps more) to build and prove out a pilot plant. Thus, reliable vitrification of plutonium mixed with high-level wastes in a new plant in the United States could not in all likelihood begin for 15 years or more. It could be completed in a few years after that, giving a total time frame of about 20 years.

Vitrification of Plutonium Alone or Mixed with Heavy Metals

Of all the options, the vitrification of plutonium alone or mixed with heavy metals that are not strong gamma emitters (or are non-radioactive) could probably be accomplished most quickly. This is because such a plant would not require the massive shielding of a completely remote operation necessary for a plant that incorporates strong gamma-emitting materials like cesium-137. Plutonium-239 is an alpha-emitter which emits relatively weak gamma rays, and hence the shielding requirements are far less stringent. However, since plutonium is highly carcinogenic once it is inside the body, appropriate precautions, such as glove boxes, are required in its processing.

The vitrification of reactor-grade plutonium that has a high americium-241 content will require greater shielding precautions than weapon-grade plutonium. As americium-241 builds up over a few decades due to the decay of plutonium-241 (half-life, 14.4 years), the glass will actually become a stronger gamma emitter. This will make vitrification of civilian plutonium with actinides or rare earths more proliferation resistant than for weapon-grade plutonium.

In the last few years, a new melter which stirs glass at several hundred revolutions per minute, causing it to foam, has been tested for its potential to vitrify radioactive and mixed wastes.[65] Because it relies on

65 Bickford 1990

active "stirring" by rotating the glass melt, the materials to be vitrified are more rapidly and thoroughly mixed in the melter. As a result, the size of the melter is reduced by a factor of seven to ten compared to the stationary joule-heated melter which is at the core of DWPF.

Stirred-glass melters come in various capacities, from small bench-scale models having a melt area of only 0.25 square feet, to melters with capacities larger than DWPF.[66,67] They can accept dry feeds or slurry feeds (like those for which DWPF is designed).

Two of these melters have been ordered for pilot-scale testing by Savannah River. One unit is slated to test vitrification of "low-level" radioactive waste. The second is to be tested as a possible replacement for the DWPF melter.[68]

Since the stirred glass melter is considerably smaller in size (for the same glass-making capacity) than the one in DWPF, it is also less expensive and can be built more rapidly. A pilot plant with a melter of comparable capacity to the DWPF (about 100 kilograms of glass per hour) could possibly be built and ready for testing in two years or less for about $10 million,[69,70] though additional time may be required for budgeting and other preparatory work. Alternatively, one of the two units already ordered could be used to begin pilot-scale tests on vitrification of plutonium.

While there is commercial experience with stirred glass melters, they have not yet been used for vitrification of plutonium. In view of our earlier discussion and analysis of past problems with vitrification in the U.S. nuclear weapons complex, we feel that gaining experience in vitrifying plutonium is vital to the eventual success of this approach. This would also allow time for other parallel work needed on environmental and long-term disposition issues, as is discussed in Chapter 6.

The small size of a stirred-glass melter presents numerous technical and policy advantages. These can be realized, especially in those cases where plutonium is vitrified, without mixing it with strong gamma-emitting materials. First, stirred-glass melters can be built relatively

66 Ray Richards, personal telephone communication, Glasstech, Inc., September 22, 1990.
67 Fact Sheets, Glasstech Stir-melter Systems, Perrysburg, Ohio. (undated)
68 John Plodinec, Savannah River Glass Technology Center, personal telephone communication, August 25, 1992.
69 John Plodinec, Savannah River Glass Technology Center, personal telephone communication, August 25, 1992.
70 Ray Richards, personal telephone communication, Glasstech, Inc., September 22, 1992.

rapidly and inexpensively, both due to their small size and because extensive shielding is not required. Small pilot plants could be built and tested within three to four years. Such pilot plant experience is vital to proving the process and ensuring that larger-scale efforts will proceed more smoothly than prior DOE vitrification efforts.

Second, the modular nature of the technology means that the plants can be built where the plutonium is located (though this is not necessary). In this way, the transportation of plutonium can be minimized. This could be a very large advantage in the case of Russian plutonium.

Third, the low cost of the plants means that a number of experimental approaches to vitrification of plutonium residues within the weapons complex could be tried. For instance, at least a portion of the plutonium nitrate solutions stored at Savannah River that pose serious safety questions could be vitrified in order to test the process for suitability at other sites in the U.S. and abroad where similar problems exist.

There are also other glass-making technologies that could be used for plutonium vitrification. One commercial technology, called "direct-induction, cold-crucible glass melting technique," has been used in metallurgical high-temperature applications and is now being tested for radioactive materials. This melter appears to possess important advantages in terms of low maintenance resulting from low crucible temperatures. A layer of solid glass separates the cooled crucible wall from the glass melt.[71]

Another technology that is not yet commercial that has potential for residues is the direct conversion melter, discussed in Chapter 4.

The operation of three to four pilot plants testing different technologies and plutonium-bearing materials would help the DOE prepare a better Environmental Impact Statement on plutonium vitrification based on actual operating experience.

Existing Facilities for Processing of Plutonium Metal to Prepare it for Vitrification

As discussed in Chapter 4, plutonium would have to be converted to nitrate form and/or oxidized prior to vitrification. There are existing facilities to accomplish both. However, the facility at Hanford, called the Plutonium Finishing Plant (PFP), is old, environmentally suspect, and

---

[71] Moncouyoux et al. 1991.

slated to be shut down permanently. It may require extensive refurbishment as well as safety and environmental studies to make the PFP suitable for the task of pre-processing plutonium, and even then it may never be on a par with newer facilities.

The plutonium conversion facilities at the Savannah River Site are of more recent design and construction, but they are also old, environmentally suspect, and slated to be shut down permanently. The old HB-line there "has recently been upgraded to prepare Pu-239 oxide from some feedstocks for storage or conversion to metal...."[72] The DOE has also stated that the modified HB-line can also recover enriched uranium.[73] The HB-line has been used mainly for oxidation of neptunium-237 and plutonium-238, the former being the raw material for the latter.[74] However, the facility has never been fully operated since the time of its upgrading.

There is also the New Special Recovery (NSR) facility at Savannah River Site whose functions are still classified but which is apparently designed to perform a wide variety of operations, including plutonium recovery from residues. It was designed to be an upgrade of older recovery facilities in the F area.[75] Conversion of plutonium to nitrate or oxide form could therefore take place at Savannah River Site. Appropriate means of transporting the plutonium to the DWPF would have to be constructed. These would not be needed if the far more compact stirred melter is built adjacent to the NSR facility. According to Plodinec, a "stirred melter could be put within the New Special Recovery facility."[76] However, plutonium cannot be mixed with calcined fission products or cesium capsules unless the NSR facility is considerably modified because it lacks adequate shielding for worker protection.[77]

Whatever plutonium disposition activities are undertaken at the Savannah River Site, there will be a need for a thorough environmental review of them so that operations are conducted with far greater thought to their environmental consequences than they have been in the past. An Environmental Impact Statement is underway at the Savannah River Site that could, in principle, address most or all of these issues.

72 ERF 1992, p. 42.
73 DOE statement quoted in ERF and NRDC 1992, p. 45.
74 ERF and NRDC 1992, p. 45.
75 ERF and NRDC 1992, pp. 42–43.
76 M.J. Plodinec, personal communication, October 12, 1992.
77 J.M. McKibben to Peter Johnson, personal communication, April 30, 1993, p. 4.

CHAPTER 6

Repository Disposal of Plutonium-Containing Waste
Some Technical Considerations

Plutonium can be disposed of in a deep geologic repository either directly (as a metal or oxide), or after processing into spent fuel or glass logs. At present there are two potential repository locations in the United States where such disposal could occur: Yucca Mountain in Nevada and the Waste Isolation Pilot Plant (WIPP) near Carlsbad, New Mexico. However, there are potentially serious concerns surrounding disposal at both of these locations.

WIPP is designed to hold transuranic waste generated by Department of Energy facilities. The waste would be disposed of in an environment that would tend to prevent its oxidation and hence its disintegration into fine particles (called a reducing chemical environment), and thereby retard its dispersal into the environment. However, there is considerable doubt whether WIPP will be permitted to open because of various technical and regulatory compliance issues. The Environmental Protection Agency (EPA) is currently conducting tests related to determining its suitability. WIPP also has very limited space and cannot accommodate all the transuranic wastes in the weapons complex. Finally, plutonium mixed with fission products could probably not be legally disposed of at the Waste Isolation Pilot Plant, which is meant only for transuranic wastes. Transuranic wastes contain high concentrations of radioactive materials like plutonium with atomic numbers greater than 92 (the atomic number of uranium), but do not have the high concentrations of fission products, like cesium-137, that characterize high-level wastes.[78]

The Yucca Mountain site in Nevada is being considered for disposal of spent fuel rods and vitrified high-level reprocessing waste from

---

[78] For a discussion of waste classification issues, see Makhijani and Saleska 1992.

military plutonium production, including the glass logs to be made at the DWPF. However, Yucca Mountain is an oxidizing environment, which would promote conversion of plutonium into fine oxide particles, making dispersal more likely. It is currently the only site being considered for high-level waste disposal in the United States.

Neither of these sites is designed for disposal of plutonium pits from dismantled nuclear weapons. Since plutonium disposed of in the form of pits would not be chemically processed and diluted, it would have to be packaged in small quantities or in special containers so as to prevent accidental criticality. Increased criticality concerns and the potential for recovery of plutonium from the repository may also present difficult questions regarding repository licensing, thus complicating the already complex problem of high-level radioactive waste disposal.

The option of burial in a geologic repository without processing was rejected by the NAS study. However, the same study did recommend that disposal of plutonium in deep boreholes be studied further as an option (see Chapter 8).[79]

Current estimates for the cost of projected geologic repository disposal of spent fuel from commercial nuclear power reactors run at about $300,000 per ton of heavy metal in spent fuel (in 1988 dollars).[80] The costs of plutonium disposal per unit weight would be higher than that for spent fuel, because, unlike spent fuel, plutonium must be highly diluted in order to make re-extraction difficult. This greatly increases the weight and volume of the material to be disposed of. Assuming a cost on the order of several million dollars per metric ton of plutonium, disposal costs would be on the order of a hundred to several hundred million dollars for repository disposal of 50 metric tons of plutonium. This assumes that DOE cost estimates for a repository are valid projections.

We will now examine some specific issues relating to repository performance that are relevant to plutonium disposition.

Repository Performance

The long half-life of plutonium-239, over 24,000 years, is one of the main challenges facing its disposal as a waste. It is not possible, at present, to predict with any level of confidence the performance of any waste form in a repository setting for periods well in excess of 100,000 years.

79 NAS 1994, pp. 187, 196–199.
80 Makhijani and Saleska 1992, pp. 67–68.

However, some investigations of glass and other waste forms provide indications as to which strategies may hold better potential for success.

Understanding the behavior of both man-made and natural materials in the repository is vital to assessing repository performance. In general, the modeling of the performance of engineered barriers and waste forms in a repository should incorporate the following:

- Data from and analysis of natural analogs in order to understand the behavior of engineered barrier systems in their assumed repository settings;
- Relevant laboratory data to complement the understanding derived from natural analogs;
- Careful theoretical analyses of canister, engineered barrier, and repository performance incorporating the first two elements;
- Geologic and hydrogeologic field data that are needed to assess repository performance.

For the present, the considerable laboratory and theoretical work on borosilicate glass indicate that its performance may be satisfactory in several geologic settings. There are potential concerns under three specific geologic circumstances:

- High ground water velocity;
- Attack by hot water vapor in unsaturated conditions (known as "hydration aging");
- Flooding of the repository, causing the boron, which has a higher leach rate than plutonium, to leach from the borosilicate glass.

High Groundwater Velocity

High ground water velocity (on the order of one meter per year or more) across the surface of borosilicate glass prevents a protective layer of chemicals from being formed on the glass surface, causing relatively rapid surface erosion. Such erosion could cause plutonium to leak out of the repository and be carried by the water into the human environment. It is important to note that a release rate of just one part of plutonium in 10,000 per year is a rather high rate, because it would lead to the release of most of the inventory of plutonium-239, since it would have decayed by only about 25 percent during that period.[81]

Hydration Aging

If a repository is hot and contains some water vapor, but is unsaturated, the hot vapor attacks the surface layers of borosilicate glass, causing them to change chemically far more rapidly than they would if the repository were either completely dry or saturated with water. Subsequent flooding of the repository causes the surface layers of the glass to disintegrate and form colloids (fine suspended particles) that are transported by water. Actinides, notably plutonium and americium, are preferentially concentrated in the colloids. As a result, there is a high potential for the colloidal radionuclides to contribute to radiation doses, if the geologic formation is such that it does not absorb them.

There is a potential for hydration aging at Yucca Mountain, which is to be designed as a dry repository (provided that construction is approved there), if conditions change and if sufficient water vapor penetrates the repository location.[82,83]

The susceptibility of borosilicate glass to hydration aging phenomena also points to the possibility that it may be easier to recover plutonium from this type of glass, compared to other glass compositions or waste forms.

Whether there will be a repository at Yucca Mountain or whether WIPP will open are matters of conjecture at the moment. Thus, while some consideration needs to be given to compatibility of waste forms with these geologic repository locations, this factor must be balanced against the increased security risks that would be inevitably associated with any delay in converting plutonium into non-weapons-usable forms.

The space requirements for plutonium canisters not mixed with high-level wastes would not be great compared to other high-level wastes that may be disposed of at Yucca Mountain. If the option of adding plutonium to high-level wastes during vitrification is chosen, then there would be no additional space requirements unless greater spacing between the canisters is needed.

81 Makhijani and Tucker 1985, pp. 52–53.
82 Makhijani 1991.
83 Bates 1992.

Space for the additional plutonium canisters may be more of a concern at WIPP, since disposal of plutonium there would have to be weighed against the fact that there is currently no room in WIPP for considerable quantities of existing transuranic wastes.[84]

Criticality Concerns

The leach rate of boron

If flooding of the repository occurs, then boron is likely to leach out of the glass faster than plutonium, raising concerns regarding criticality. DOE extends this criticality concern for over a billion years since plutonium-239 decays into uranium-235, which is also a fissile material. (Uranium-235 has a half-life of over 700 million years.) Therefore, the best precaution is to ensure that the leach rate of plutonium or uranium is comparable to or slower than the leach rate of the neutron absorber.

Neutron absorbers (also called "neutron poisons") whose leach rates are closer to the leach rate of plutonium could be added to address criticality concerns.[85] One such neutron absorber may be lanthanum, which is a metal in the rare earth series. These metals have similar chemical properties to another, heavier series of elements known as actinides. Both plutonium and uranium belong to the actinide series of elements. As discussed in Chapter 4, it would be advantageous from a non-proliferation standpoint to mix plutonium either with uranium (from the actinide series) or with a rare earth element such as europium or gadolinium.[86]

Most repositories proposed for hard rock locations would be saturated—that is under anticipated conditions of disposal, water would flood the repository well before the radioactivity in the wastes had decayed away. Criticality concerns are naturally higher at such locations because water is a neutron moderator; its presence therefore lowers the amount of plutonium needed to form a critical mass. However, even in locations such as Yucca Mountain, where the proposed repository is designed to be in a location that is now dry, the anticipated duration of the criticality potential is far longer than the periods for which the geologic environment might be expected to remain stable. This is due to the very long half-life of uranium-235 (see above).

---

84 Makhijani and Saleska 1992, p. 58.
85 Westinghouse 1993, p. 9.
86 von Hippel et al. 1993, p. 49.

Plutonium-239 Decay

Plutonium-239 decays into uranium-235. In a repository environment uranium may concentrate in the same way that it does in natural uranium deposits. In some cases, quite high concentrations of uranium can occur. In normal circumstances, criticalities do not occur in ore because the concentration of uranium-235 in natural uranium is low. However, about three billion years ago, when the isotopic concentration of uranium-235 in nature was much higher (it has changed due to faster decay of uranium-235 relative to uranium-238) a sustained criticality occurred in nature in what is now the country of Gabon in West Africa.[87]

Disposal of plutonium that has been diluted only with fission products or not at all has the potential that the uranium-235 could concentrate and cause a criticality. This concern can be substantially reduced if the plutonium is diluted with a sufficient quantity of depleted uranium (which is essentially all uranium-238) and/or a rare earth element. Analysis of the potential of disposal locations for such a criticality could also help alleviate this concern.

Sub-Seabed Disposal

Sub-seabed disposal is a potential alternative to repository disposal of high-level radioactive waste. Preliminary investigation of sub-seabed clay deposits in areas which appear to have long-term stability indicated that the chemistry of these deposits may be suitable for the isolation of nuclear wastes. One of the main advantages of direct disposal of plutonium in sub-seabed sediments is the great difficulty in retrieving it, though future technological developments may change that. The NAS study recommended further research on this option for excess plutonium disposal.[88]

Like all other options for plutonium disposition, this one poses many potential problems. Among them are transportation accidents, the difficulty of adequate site characterization, and obtaining international political agreement. In view of the reality that all options have serious drawbacks associated with them, a modest level of research is justified. However, DOE has terminated research on sub-seabed disposal for high-level waste in the U.S., despite a Congressional mandate to DOE to conduct such research.[89]

87 LaMarsh 1983, p. 181.
88 NAS 1994, pp. 199–202
89 For more information on sub-seabed disposal, see OTA 1986.

CHAPTER 7

Highly Enriched Uranium Disposition
Technical Aspects

Both plutonium and HEU pose security and environmental risks, but in somewhat different ways. HEU is about 1,000 times less radioactive per unit weight than plutonium-239, but it is easier to make HEU into a bomb by using the less difficult "gun-type" of design, such as the one used in the weapon dropped on Hiroshima.[90] The disposition of HEU is, in principle, considered a simpler problem than plutonium disposition for two reasons. First, the technology to make it non-weapon-usable exists; second, the resulting product, low enriched uranium, has a well-established commercial application of use as a fuel in civilian nuclear power plants.

Issues concerning the disposition of HEU and plutonium are related in various ways. Safeguards and materials accounting issues are essentially the same. Another connecting thread is that the use of LEU in nuclear reactors creates more plutonium, though this is a general characteristic and not one specific to LEU derived from HEU. Another is that LEU made from HEU can be used in reactors in place of MOX fuel or in addition to it.

Weapon-grade uranium, which contains 90 percent or more of uranium-235, is used in nuclear weapons and as fuel for naval and research reactors.[91] Unlike plutonium, most of the HEU is in military stocks. (Most plutonium is in spent fuel from nuclear power plants). The major

---

90 A gun-type device involves two sub-critical masses of uranium brought together by the use of a conventional explosive to form a supercritical mass.

91 Uranium with an isotopic composition of 20percent or more uranium-235 is classified as highly enriched.

civilian use of HEU is in research reactors. Both naval and research reactors can use LEU as fuel, though their operating characteristics would be somewhat different as a result.

Worldwide HEU Inventory

The worldwide inventory of highly enriched uranium is not well known, since information about the production of this material is still largely classified, notably in Russia. The United States has only recently become an exception. The biggest producers of HEU have been the United States and the former Soviet Union. The total world production of HEU appears to have been about 2,300 metric tons, but there are considerable uncertainties in this figure, deriving mainly from uncertainties about Russian production (see below).

United States

The DOE has recently declassified the amount of HEU the United States produced between 1945 and 1992. The total production in this period was 994 metric tons.[92] There has been no HEU production in the U.S. since 1992; therefore the figure of 994 metric tons represents the cumulative current production to date. However, the amounts incorporated into weapons and already used in naval reactors have not yet been declassified. Albright et al. have estimated that the amount currently in the military stockpile, including that in weapons, is about 550 metric tons; this estimate is based on a total production estimate of 720 metric tons.[93] These two estimates imply a past use of 170 metric tons.

The 1993 RAND study estimated that in the year 2003, after the dismantling of 14,000 nuclear warheads, the consumption of 48 metric tons of HEU in naval reactors, and retention of 88 metric tons of HEU as a military stockpile, the surplus of HEU in the United States will be about 339 metric tons.[94] However, this estimate assumes that there was 550 metric tons in military (stockpile and warhead) use as of 1991, with an implicit acceptance of the overall production estimate of Albright et al. of 720 metric tons.[95] Given that the actual production figure was 994 metric tons, the surplus could be greater.

92 DOE 1994, p. 52.
93 Albright et al. 1993, pp. 49–50.
94 Chow and Solomon 1993, pp. 11–12.
95 Albright et al. 1993, p. 50.

Highly Enriched Uranium Disposition

If one assumes that the RAND figures for future requirements are plausible and that the estimates of Albright et al. of past use are also accurate, then the surplus HEU in the Unites States by 2003 could be over 600 metric tons, which is much higher than that projected by RAND. Further arms reduction agreements would increase this total. As of December 1994, the United States government does not have active plans to blend down a substantial portion of this surplus into LEU for commercial use as a fuel.

Russia, France, China, and Britain

The amount of HEU produced by the Soviet Union has not been declassified. The 1993 RAND study, citing an estimated inventory in 1991 of 720 metric tons of HEU in warheads and in military stockpiles in the former Soviet Union, calculated that by the year 2003 the surplus of HEU would amount to 637 metric tons.[96] The Russian Minister for Atomic Energy, Viktor Mikhailov, has cited a figure of 1,250 metric tons for HEU production in the former Soviet Union.[97] On this basis, the Russian surplus of HEU could be far larger than the 500 metric tons it has agreed to sell to the United States, and also considerably larger than the 637 metric tons estimated by RAND.

Albright et al. have estimated that the combined holdings of HEU by France, Britain and China to be in the range of 25 to 55 metric tons.[98]

Blending Down of HEU for Use in Light Water Reactors

The principal advantages of blending down HEU into LEU are:

- It is practically impossible to manufacture a weapon using LEU, unless the LEU is re-enriched.

- LEU has an economic value as a nuclear fuel so long as there are operating nuclear power plants.

---

96 Chow and Solomon 1993, p. 12. The estimate of 720 metric tons of HEU in the former Soviet Union is from Albright et al. 1993, p. 198.

97 NAS 1994, p. 131. Mr. Mikhailov has also been quoted a saying that 500 metric tons represents 30 to 40 percent of Russia's HEU supply. This would put the upper limit of Russian HEU at 1,667 metric tons. Helen Hunt, personal communication, November 8, 1994, citing Nukem Market Report, October 1993, p. 25. It is difficult to say whether such comments are meant as information regarding Russian stocks or whether they have other purposes.

98 Albright et al. 1993, Table 4.1 page 48.

- The use of LEU could displace uranium produced from uranium mines, thereby reducing the generation of mine wastes and mill tailings.
- The energy required for uranium enrichment is eliminated (since it has already been expended in producing HEU).
- Uranium metal is a very reactive element; its conversion into uranium dioxide (the chemical form used in most reactors) makes it much less hazardous to store.
- Safeguards requirements for LEU are lower than those for HEU if there are no operating uranium enrichment facilities, particularly those that use centrifuge enrichment technology.
- A portion of the LEU made by blending down HEU could be used as a strategic reserve in order to discourage further civilian plutonium separation (see below).

One potential disadvantage is that the conversion from HEU to LEU on the scale required could take decades. A second problem is that the large scale and long duration of the blending down operation will increase the risks of diversion of HEU. It is to be noted that the blending down associated with the U.S.-Russian HEU purchase agreement will likely be done in Russia. Another disadvantage is that the use of LEU in reactors results in the creation of plutonium. However, as noted above, this is not a problem peculiar to the use of LEU made by blending down HEU. Rather, this is a more fundamental question concerning the relationship between the use of nuclear power plants and future security issues.

Blending Down HEU into LEU[99]

The blending down of HEU to LEU can theoretically be done by mixing the HEU with natural uranium (0.711 percent uranium-235), depleted uranium (in the range of 0.2 to 0.4 percent uranium-235), or slightly enriched uranium (about 0.8 percent to 2 percent uranium-235). The amount of blending material (called blend-stock or matrix) required

---

99 Much of the information presented in this section is based on a paper by Norman E. Brandon of Nuclear Fuel Services, Inc. Unless otherwise stated, Brandon 1993 (see reference list) is the reference for this section on blending down techniques.

depends on which of these three options is chosen. This also affects the amount and isotopic composition of the final product, notably the amount of uranium-234 in it. Uranium-234 is a trace isotope in uranium, but it is important because it has a far higher specific activity (radioactivity per unit weight) than uranium-238 or uranium-235, so much so that it is the main determinant of the specific activity of HEU. The blend-stock could be in metal, oxide, or hexafluoride form.

The amount of LEU that will result from blending down HEU will be very substantial. The total amount of final product will be the sum of the blend amounts.[100] Table 5 shows the amounts of final product of 4.4 percent LEU that would be obtained using 0.2 percent depleted uranium, natural uranium, and 1.5 percent blend-stock, and 500 metric tons of 93.5 percent HEU. We have chosen the figure of 500 metric tons, since that is the amount that the United States has contracted to purchase from Russia.

It is important to note that the most slightly enriched uranium that may be used for blending down HEU in Russia is likely to originate from the reprocessing of spent fuel. This uranium contains uranium-236, which is an undesirable isotope because it makes the resulting fuel more radioactive (see below).

Table 5 shows that the U.S. purchase of 500 metric tons of HEU from Russia will result in large quantities of LEU, no matter what blend-stock is used. The Russian proposal is to use 1.5 percent enriched uranium as blend-stock, which would create the largest amount of LEU, about 15,900 metric tons of 4.4 percent enriched uranium.[101]

100 The mathematics underlying the calculations in Table 5 is straightforward. Consider one kilogram of HEU enriched to h% that is to be blended down to a% LEU. Let the blend-stock content of uranium-235 be b%. The problem is to find the amount of blend-stock required to get a final product of a% LEU. Let the blend-stock amount be x kilograms. The total content of uranium-235 in the input materials is $(0.01*h + 0.01*b*x)$. The final enrichment is a% and the final amount of product is x+1 kilograms. The final amount of uranium-235 is therefore $0.01*a*(x+1)$ kilograms. The final and initial amounts of uranium-235 must be equal, so that the equation for determining the amount of blend-stock is:

$$(0.01*h + 0.01*b*x) = 0.01*a*(x+1)$$

Solving for x, we get, $x = (h-a)/(a-b)$

The total final product, f, is the sum of the initial 1 kilogram of HEU and the x kilograms of blend-stock, i.e. $f = x+1$. If the initial amount of HEU is S kilograms, the blend-stock required and the amount of final product are both S times greater than the corresponding quantities for 1 kilogram of HEU.

101 For a given enrichment of HEU, the amount of LEU product depends on both the uranium-235 content of the blend-stock and the final enrichment.

TABLE 5. **Amount of Blend-Stock and Final Product for Blending Down 500 Metric Tons of 93.5% HEU**

BLEND-STOCK	HEU quantity, m.t.	BLEND stock, m.t.	4.4% LEU product, m.t.
Depleted uranium (0.2% U-235)	500	10,600	11,100
Natural uranium (.711% U-235)	500	12,100	12,600
Slightly enriched uranium (1.5% U-235)	500	15,400	15,900

Note: All figures rounded to the nearest 100 metric tons. Metric tons is abbreviated as m.t..

So far the United States has committed to blending down a very small amount of HEU from its inventory. As a result the United States is holding on to essentially all of its HEU and has not declared it a surplus. This position is in contrast to the Russian readiness to sell a substantial fraction of its stock for blending down into LEU. This lack of reciprocity in U.S. HEU policy might provoke a backlash in Russia, especially in a climate of growing Russian nationalism.

The reductions in naval nuclear weapons under current U.S. arms reduction commitments call into question whether the old policies of heavy reliance on naval nuclear reactors should be continued in the post-Cold War period. The United States needs to more carefully evaluate the proliferation risks posed by a policy of indefinite storage of HEU, especially in light of the fact that Russia is likely to maintain a large HEU reserve even after the blending down of 500 metric tons that it has sold to the U.S. is fully implemented.[102]

As we have discussed, the 1993 RAND report estimates that almost 1,000 metric tons of HEU are likely to become surplus by the year 2003 under the assumptions of nuclear weapons reduction programs in place as of 1993. This would mean that a total of about 22,000 to 32,000 metric tons of 4.4 percent LEU could be manufactured, depending on the

---

102 For discussion of the reciprocity issue, see OTA 1993, pp. 101–102.

blend-stock that is used. Further arms reduction agreements could increase this total.

The U.S. purchase of Russian HEU blended down to 4.4 percent LEU would supply the entire U.S. requirement for LEU for almost 8 years at current rates of consumption.[103] The actual duration for which the LEU supply lasts depends on the pace at which existing nuclear power plants are retired, the average electricity output of existing plants in the future, and whether and how many new nuclear power plants will be built. If the entire 1,000 metric tons of surplus HEU estimated by RAND (U.S. and Russia, combined total) is blended down, the resulting LEU would suffice for over 15 years of U.S. LEU requirements at current consumption levels. As discussed above, the actual HEU surpluses may be even higher.

There are a number of practical issues associated with the conversion of HEU into LEU, and there are related policy consequences. We will first discuss the methods that can be used to blend down HEU, then move on to a discussion of some areas of caution and concern.

The form of HEU in nuclear weapons is uranium metal. HEU can be blended down as a metal by first melting it and then mixing it with molten depleted, natural, or slightly enriched uranium metal. It can also be further chemically processed into uranium oxide, uranyl nitrate (a liquid), or uranium hexafluoride and then blended. There are therefore many possible approaches to carrying out the blending down of HEU. Figure 5 shows a flow-chart of possible blending down approaches.

Blending down operations must produce reactor fuel that meets U.S. specifications if it is to be used in the U.S. without further processing. U.S. specifications limit the concentrations of uranium-236 and uranium-234 in fuel; both of these are unwanted isotopes. Uranium-236 is a neutron absorber which is formed during irradiation in a nuclear reactor and is very hard to separate since its atomic weight is very close to the atomic weight of uranium-235. Its presence in LEU fuel would lower fuel performance. Uranium-234 increases worker hazards in fuel fabrication. The American Society for Testing Materials has specification limits for the concentration of uranium-234 and uranium-236 in fuel.

---

103 U.S. consumption of natural uranium in the form of U_3O_8 has been about 20,000 metric tons per year in recent years, which is equivalent to about 17,000 metric tons of elemental uranium. It takes about 8.22 kilograms of natural uranium to produce one kilogram of 4.4 percent enriched uranium. This means that 15,900 metric tons of 4.4 percent uranium (elemental basis) obtained from Russia would be equivalent to about 131,000 metric tons of natural uranium, or about 7.7 years consumption.

The concentration limit for uranium-234 is 10,000 parts per million parts of uranium-235, the concentration for uranium-236 is 5,000 parts per million parts of uranium-235.[104]

Blending Down of Uranium Metal

Uranium metal can be melted and then mixed in a homogeneous fashion with molten depleted or slightly enriched uranium metal.

Blending Down Uranium in Oxide Form

HEU metal can be oxidized in air, by roasting for instance, to produce uranium oxide (U_3O_8). The powdered HEU oxide can then be mixed with yellowcake (which is produced from uranium ore and which is principally U_3O_8) to obtain the desired enrichment. The blended down U_3O_8 must then be converted to uranium dioxide, which is then fabricated into ceramic fuel pellets suitable for use as fuel in nuclear power reactors.

A second method is to process the HEU and blend-stock separately and then mix them. In this case one would first dissolve HEU metal or U_3O_8 in nitric acid and then process it further to obtain uranium dioxide (see below). The highly enriched dioxide in powder form can then be blended down with natural or depleted uranium dioxide in powder form.

Blending Down of Uranium in Liquid Form

One can also dissolve uranium metal in nitric acid. As with dissolution of uranium oxide, this yields uranyl nitrate hexahydrate (abbreviated as UNH). The solution is then purified through solvent extraction and converted to uranium trioxide (UO_3) by the application of heat.

$$UO_2(NO_3)_2 + \text{heat} \rightarrow UO_3 + 2\,NO_2 + 1/2\,O_2$$

Uranium trioxide, which is orange in color, is then reduced to uranium dioxide (a brown powder) with hydrogen gas.

$$UO_3 + H_2 \rightarrow UO_2 + H_2O$$

The uranium dioxide HEU powder can then be blended with uranium dioxide LEU powder.

---

104 Brandon 1993, p. 9.

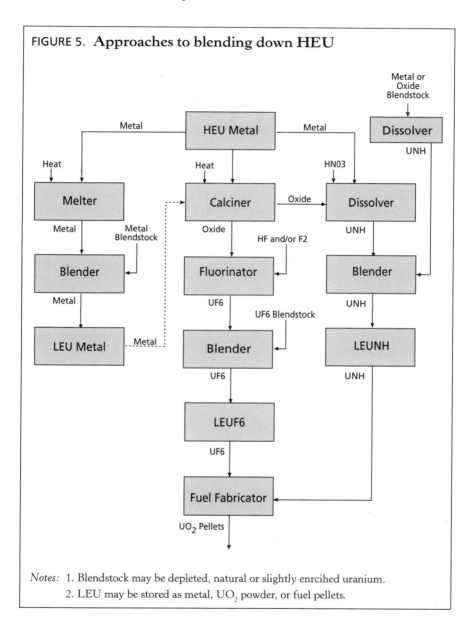

FIGURE 5. Approaches to blending down HEU

Notes: 1. Blendstock may be depleted, natural or slightly enriched uranium.
2. LEU may be stored as metal, UO_2 powder, or fuel pellets.

The methods of blending HEU and blend-stock oxides in powder form may not result in a product that is sufficiently homogenous. Therefore, any approach that uses this step may not be suitable for producing LEU fuel.

Blending Down of Uranium in Gaseous Form

Another method for blending down HEU is to put both HEU and LEU first into the form of uranium hexafluoride (UF_6). One advantage of using this method is that fuel fabricators already have the facilities for converting UF_6 into uranium dioxide fuel pellets.

One way to produce uranium hexafluoride is to first convert uranium metal into uranium dioxide as described above. The uranium dioxide is converted to uranium tetrafluoride by reaction with hydrofluoric acid:

$$UO_2 + 4\ HF \rightarrow UF_4 + 2\ H_2O$$

Uranium tetrafluoride is then converted to uranium hexafluoride with fluorine gas:

$$UF_4 + F_2 \rightarrow UF_6$$

Other intermediate oxides may also be used to produce UF_6. The highly enriched UF_6 can be then diluted with a blend-stock of depleted, natural, or slightly enriched UF_6.

Blending Down of HEU in the United States

HEU exists in several forms in the U.S., including metal and UF_6. How much blending down work is done in the United States depends in part on whether any Russian HEU is blended down here. A second unknown is how much of its own HEU the U.S. will eventually blend down.

The United States has some physical capacity to blend down HEU in the Y-12 plant at Oak Ridge. According to the 1993 OTA report, this capacity is limited, however, and would need to be expanded if Y-12 is "to handle adequately the much larger volumes of Russian HEU."[105] However, a careful safety, environmental, and security review would be needed before one could be sure that the plant would be able to operate in compliance with health, safety, and environmental protection laws.

In May 1993, Nuclear Fuel Services of Erwin, Tennessee was granted a license amendment by the U.S. Nuclear Regulatory Commission (NRC) to blend down HEU to LEU as uranyl nitrate.[106,107]

---

105 OTA 1993, p. 104.

106 Norman Brandon, personal fax communication to Arjun Makhijani, November 7, 1994.

107 One should note in this context that, in at least one instance, the State of Tennessee has judged Nuclear Fuel Services to be unsuitable for storing 750 barrels of plutonium-contaminated mixed waste, and asked that the wastes be shipped to Oak Ridge. *Knoxville News-Sentinel*, Nov. 22, 1990 and December 12, 1990.

Highly Enriched Uranium Disposition

The DOE's plans as of late 1994 call for down-blending using UF_6 at the Portsmouth Gaseous Diffusion Plant in Ohio, but no date for that work has been set. Highly enriched uranium hexafluoride, which is presently stored as a solid in 5-inch diameter by 36-inch tall containers, would be heated to a gas to be fed into the plant. There, it would be mixed with a matrix of uranium hexafluoride containing 1 percent uranium-235 to produce low enriched uranium hexafluoride to be used, after conversion to uranium dioxide, in light water reactors.[108]

While there is some blending down capacity in the U.S., there is at present no installed capacity to convert HEU oxide or metal into UF_6. Nuclear Fuel Services, at its plant in Erwin, Tennessee, is currently designing the necessary facilities for the conversion of HEU metal into hexafluoride, followed by purification, and blending. These facilities are intended to handle Russian HEU. The capacity would be 10 metric tons of HEU per year, with modular designs to increase the capacity if needed.[109]

Blending Down in Russia

The process that would be used to blend down HEU in Russia has not been made public. However the main outline of the process is thought to be the following: The metal would first be oxidized to U_3O_8 using a wet chemical process and then converted to UF_6 before final blending down. The product, which would contain 4.4 percent uranium-235, would then be loaded into cylinders for shipment to the United States.[110]

Vitrification of HEU

Vitrification of HEU could be an interim, perhaps partial, alternative to blending it down to LEU. The main objective would be to put HEU as rapidly as compatible with safety, health, and environmental protection into a form not usable in weapons without complex processing. Since HEU presently has commercial value as a potential nuclear reactor fuel, it would probably not be acceptable to owners of HEU (the governments of Russia and the United States) to vitrify HEU mixed with

---

108 Norton Haberman, Acting Director, Office of Uranium Programs, Office of Nuclear Energy, DOE, personal fax communication to Annie Makhijani, Sept. 27, 1994.

109 Brandon 1993, p. 9.

110 Norton Haberman, Acting Director, Office of Uranium Programs, Office of Nuclear Energy, DOE, personal fax communication to Annie Makhijani, Sept. 27, 1994.

fission products. The re-extraction costs in that case would be very high—probably higher than LEU production from newly mined uranium ore. Further, since such vitrification would take a very long time, it would require prolonged storage of HEU, and present no advantages over blending down HEU.

HEU could also be vitrified in the near to medium-term either alone or with non-radioactive materials, along the lines that we have already discussed for plutonium. Such vitrification of contaminated HEU might be considered first, since such HEU would be impractical to fabricate into fuel, given that cleaner HEU is in considerable surplus.

The vitrification option for HEU has not been given extensive serious attention because LEU has evident economic value so long as there are nuclear power reactors in operation. However, the time-table for blending down HEU is at present so long, that vitrification should be more carefully evaluated as a means to store HEU in non-weapons-usable form and hence decrease proliferation threats. It is not clear whether the economics of recovery would be so adversely affected as to make LEU made from mined uranium more economical. This issue needs careful evaluation. One way to view the cost would be to consider it as part of the effort to reduce the security costs of having made HEU at all. In the alternative, a large increase in blending down capacity and capacity for storage of LEU would be needed.

Hazards of Uranium Processing and Storage

The storage of uranium metal regardless of its degree of enrichment poses significant hazards due to its chemical reactivity. Uranium metal reacts with the moisture to form pyrophoric surfaces. These reactions result in uranium dioxide (UO_2) and uranium hydride (UH_3), which cause the metal to swell and disintegrate.[111] These hazards are exacerbated when uranium metal is finely divided because uranium metal powder or chips can spontaneously ignite at room temperature. In contrast, during the storage of massive uranium metal (that is metal in large chunks), a thin protective oxide film forms on the surface, reducing the danger of spontaneous ignition.

111 UEO 1990, p. 36.

Highly Enriched Uranium Disposition

Regardless of the degree of blending down and the nature of the blendstock, there are precautions that must be taken in the blending down process:

- Accidental criticality must be prevented.
- Accidental leakage of uranium or the hazardous chemicals used to process it must be prevented.
- Routine discharges of radioactivity to the air, water and soil must be minimized.
- Worker health must be protected.

Criticality Concerns

The storage of HEU poses dangers of criticality if the uranium parts are too close together. The uranium parts have to be spaced in such a way that there is no danger of creating an accidental critical mass. A recent inspection found inadequate storage of HEU from disassembled A-bombs at the Y-12 plant in Oak Ridge, Tennessee. The uranium metal parts were stored closer than safety rules permit.[112]

Blending down HEU also poses criticality concerns, especially in the early stages of the blending process. Criticality control is exercised to ensure that the mass in any process or storage vessel at any time does not exceed the critical mass for a given form and geometry of uranium. To do this, equipment must be designed in such a way as to ensure that not more than 10 kg of HEU accumulates in any one location. Columns, piping, holding tanks, and blending tanks therefore need to be of the appropriate size.[113]

Environmental Issues

The greatest single potential environmental advantage of blending down HEU into LEU is that it can displace the uranium mining, milling, and enrichment activities that are needed to make uranium fuel from ore.[114] Both the HEU and the blend-stocks have already been produced—in other words, the uranium has already been separated from the ore and

---

112 *New York Times*, October 4, 1994.
113 Chow and Solomon 1993, pp. 80–81.
114 OTA 1993, p. 101.

from the other radioactive elements that are contaminants in the ore. Notable among these are heavy metals such as molybdenum, as well as thorium-230 and radium-226, both of which are decay products of uranium-238. They are present in significant quantities in mill tailings. Uranium mill tailings represent well over 90 percent of the volume of radioactive waste from uranium processing and use, though the radioactivity is primarily contained in the form of fission products in the spent fuel discharged from reactors. In addition, uranium enrichment is energy intensive, particularly in the United states, where the gaseous diffusion method is used. A far smaller quantity of energy is required for the blending down process. The energy required to enrich uranium has already been expended and incorporated (as it were) into the HEU. Therefore, there would be considerable environmental benefits from reduction in energy use in uranium processing and enrichment, provided the LEU made from HEU is actually used to displace normal commercial uranium enrichment activities.

The realization of these environmental advantages is premised on a number of assumptions:

- The blending down of uranium will be done with strict attention to minimizing releases of uranium and hazardous materials to the environment and to protecting worker health.

- The LEU produced will actually be released into the market and not held back either for strategic reasons (see below) or because of pressures to keep the mines open because of jobs. LEU might also be held back from the market due to corporate pressures for and interests in higher uranium prices.

- The LEU will be produced sufficiently rapidly and in sufficient quantity to justify closure of a significant number of mines and other downstream operations.

- Nuclear power plants will continue to operate for at least the next two decades in numbers of the same order of magnitude as those in operation today.

These environmental advantages at the front end of the nuclear fuel cycle cannot negate the fact that all use of LEU in nuclear power plants creates highly radioactive spent fuel—which is a huge unresolved problem for management and disposal. As we have noted, this problem is inherent in the use of nuclear power and is not associated solely with the use of LEU made by blending down HEU.

Diversion of HEU

HEU is a radioactive material that emits mainly alpha radiation, which is not penetrating radiation. If HEU is minimally shielded, it is difficult to detect it without elaborate instrumentation. Once stolen, HEU can be transported without immediate danger of large radiation doses to the carrier, since it is dangerous mainly if it is ingested, inhaled, or absorbed (for instance through a cut). The danger of diversion is greater in the former Soviet Union than it is in the United States. This was dramatically demonstrated by the U.S. government's "Project Sapphire" in which the U.S. government purchased about half a metric ton of HEU, reported to be poorly safeguarded, from a uranium fuel fabrication plant in Kazakhstan, and transported it to the United States.[115]

Even in the United States there has been a long-running controversy about whether about 200 pounds of HEU was at diverted to Israel from the fuel fabrication plant at Apollo, near Pittsburgh. According to journalist Seymour Hersh, the diversion did not occur; the missing HEU has been discharged into wastes, been absorbed by concrete floors, and emitted to the air and water, but it took three decades and the dismantlement of the plant to come to that conclusion.[116]

A lesser diversion from Russia, where materials accounting problems are likely to be greater than in the United States especially as regards materials discharged into waste streams, would be difficult to detect. The difficulty of detection would be greatly increased if the diversion occurred with collaboration of officials within the fuel fabrication or blending-down plant. Diversion threats also exist with storage, but they are increased with handling and processing since materials accounting uncertainties increase and the opportunities for diversion increase with the amount of handling and transportation.

---

115 *New York Times*, November 23, 1994.
116 Hersh 1991, Chapter 18.

CHAPTER 8

Policy Issues

Short- and Medium-Term Issues—Plutonium Disposition

No country has succeeded in opening a geologic repository for high-level waste disposal. Therefore, plutonium, whether in separated form, in vitrified glass logs, in spent fuel, or in residues will have to be stored retrievably for considerable periods. Moreover, even a single repository anywhere in the world is two or more decades away from opening.

The difficulties of disposition of surplus plutonium from dismantled nuclear weapons are compounded by continued reprocessing of civilian spent fuel in Russia, France, Japan, Britain, and India. The governments of these countries are wedded to civilian plutonium separation as an important long-term component of energy programs. They are very unlikely in the near-term to give up these programs unless their energy concerns are addressed. Yet, if reprocessing, whether military or civilian, continues, disposition decisions on U.S. surplus military plutonium alone will not fundamentally change the global security picture. The separation and circulation of civilian plutonium will, in the coming decades, far exceed the approximately 250 metric tons of military plutonium in the world. Moreover, reprocessing civilian spent fuel is continuing in Russia; until it is halted, the security concerns in relation to weapons-usable materials associated with the state of the economy and society there cannot be resolved. Therefore, policies directed at achieving a *universal but interim halt* to reprocessing are essential so that the plutonium problem is not being aggravated while long-term energy and security issues are sorted out.

No country now engaged in civilian plutonium production is likely to stop even on an interim basis without vigorous U.S. leadership. A clear and formal declaration by the U.S. government that plutonium is a security, environmental, and economic liability should be the starting point of such leadership. The text of a letter sent by 43 organizations and individuals to President Clinton on October 19, 1994 requesting such a declaration is attached as Appendix A to this report.

As we discussed in Chapter 3, the U.S., were it on its own, could more freely consider pursuing the MOX option for putting excess military plutonium in a proliferation-resistant form. However, the main threat over the next many years does not come from excess U.S. military plutonium, but from the situation in the former Soviet Union. Thus, in our analysis, a MOX option should be ruled out for the U.S. so that it can play the leading role that is needed to stop civilian reprocessing as well as military plutonium production throughout the world.

The only other option that has a chance of accomplishing the immobilization of plutonium into a proliferation-resistant form within a reasonable time-frame is vitrification. As discussed in Chapters 4 and 5, there are four broad technical options for plutonium vitrification:

- Vitrification with radioactive wastes. This would cause the glass logs to approximate the spent fuel standard;
- Vitrification of plutonium alone;
- Vitrification of plutonium with non-radioactive heavy metals;
- Vitrification of plutonium with depleted uranium or other similar radioactive actinide that is not an intense gamma emitter.

Plutonium processed by the last three options would not meet the spent fuel standard, but, as we have discussed in Chapter 4, the third and fourth options could approximate it on most counts, depending on the material(s) chosen as additives. The last two options may be combined.

It has generally been assumed that the spent fuel standard should be adopted for plutonium disposition probably because it is the most attractive according to certain non-proliferation criteria, in particular, the difficulty of re-extraction of plutonium and the resistance to theft that unshielded spent fuel provides.[117] It is also the strictest practical standard since civilian spent fuel has a large amount of plutonium in it. Processing excess military plutonium to a more stringent standard of re-extraction is therefore seen as a waste of money, given that plutonium

---

117 We have inserted the qualification "probably" in this sentence because mixing plutonium with materials other than fission products has only begun to be investigated in a preliminary way, and so our knowledge of the possibilities is still very limited.

could be extracted from civilian spent fuel. It is also generally assumed that mixing plutonium and fission products in the body of the glass is the way that the spent fuel standard should be achieved, if vitrification is the chosen disposition option.

These assumptions need to be refined for a number of reasons. First, gamma-emitting fission products, notably cesium-137, that give spent fuel its proliferation resistance, have far shorter half-lives than plutonium. (The half-life of cesium-137 is about 30 years compared to 24,000 years for plutonium-239.) Therefore, plutonium processed according to the spent fuel standard becomes less resistant to proliferation over time. In the course of a few hundred years, it will come to resemble vitrification of plutonium alone, which is the least strict of the vitrification options we have discussed in this book. Vitrification of plutonium with a non-radioactive chemical analog would provide a somewhat lower but much more durable level of proliferation resistance. The same is true of vitrification with uranium-238 or thorium-232, since both these isotopes have far longer half-lives than the plutonium isotopes in civilian or military plutonium.

Second, vitrifying plutonium mixed with fission products in the U.S. is likely to take long, since existing vitrification plants may be unsuitable for this purpose, as we have already discussed. Therefore, this option is unlikely to be accomplished as rapidly as would be desirable for non-proliferation reasons.

Finally, the spent fuel standard possesses a political defect if it is accomplished by mixing fission products into the glass. It is highly unlikely to be accepted by the countries that have spent and are spending large amounts of money for reprocessing civilian spent fuel. Even if the United States goes ahead and vitrifies its plutonium to this standard by mixing it with fission products, it is unlikely to persuade Russia, France, Japan, Britain, and India to do likewise. A lower level of re-extraction cost may be necessary to persuade these countries to halt reprocessing on an interim basis. Another way of stating this problem is that the spent fuel standard is irrelevant at the governmental level in countries that are now reprocessing. Putting some plutonium into spent fuel or vitrified glass would reduce proliferation threats only for a brief period if plutonium separation continues. The challenge, therefore, is to find a plutonium disposition option that will provide as high a resistance to theft as spent fuel for sub-national groups, and also pose great challenges for plutonium re-extraction for the same groups. So far as countries that now reprocess or that own separated plutonium are concerned, the main tasks are to persuade them to stop reprocessing and to ensure and verify that already separated plutonium is not used to make nuclear weapons.

These goals can be accomplished with the appropriate policies. Even countries such as Russia and Japan that are vigorous proponents of civilian reprocessing recognize four things, even if they do not often do so publicly:

- The use of plutonium in either thermal reactors or breeder reactors is not economical today without large government subsidies and it is unlikely to be economical for a several decades. They look to plutonium as a very long-term energy resource.

- All separated plutonium represents a potential security threat.

- Surpluses of both military and civilian plutonium exist and separation of civilian plutonium will increase these surpluses at least for the next couple of decades.

- A high level of international cooperation is necessary to reduce the security threat from plutonium.

Given this common ground, it may be advantageous to consider plutonium vitrification options where the level of effort of re-extraction is somewhat lower than the spent fuel standard for governments that are reprocessing today, both in terms of the expenditure and time, but still very high for sub-national groups. Evidently, this means that there is a corresponding decrease in the technical barrier to reuse by governments. This problem can be mitigated by safeguards and verification measures, which are in any case necessary for civilian and military separated plutonium. These measures should be buttressed by a multilateral agreement that plutonium, once declared surplus to national security, will never be used in weapons.

The barrier to theft of plutonium and hence to use by sub-national groups can be made high by making the canister containing the vitrified plutonium highly radioactive. In fact the level of resistance to theft provided by such canisters would be comparable to unshielded spent fuel ready for dry storage and far higher than that of spent fuel stored shielded casks such as those that are used for spent fuel transportation. Such casks are now under consideration in the United States for all spent fuel. The technical level of difficulty for re-extraction would be relatively high for sub-national groups, especially since remote handling would be required to remove the plutonium-containing glass from the highly radioactive canister. Beyond this step, vitrification with actinides or rare earths would provide an intermediate level of difficulty of re-extraction.

This complex of measures would allow governments that own plutonium today to recover it in the future, but make it very difficult for sub-national groups to do so even if diversion of the glass logs occurred. Thus, it would be less difficult to persuade governments that still see plutonium as a long-term energy asset that all excess plutonium, including civilian separated plutonium, should be vitrified now to reduce security risks, while keeping open the option of using it in the future should the need arise.

Our reasoning is similar to that which the NAS used in recommending further work on the deep borehole disposition option for excess military plutonium. The deep boreholes in which plutonium would be disposed of would be 2,000 to 4,000 meters deep. Plutonium emplaced at such depths would be far less accessible than that disposed of in geologic repositories, for which typical proposed depths are up to about 1,000 meters. The NAS study recommended further research on this option as a possible alternative to vitrification of plutonium and/or use of MOX fuel, even though it does not meet the spent fuel standard. This is because deep borehole is a disposition option that presents an intermediate level of difficulty of recovery of plutonium for governments but a high-level of difficulty for sub-national groups. According to the NAS, this potential for recovery may be an advantage with respect to governments, like Russia, that believe plutonium may one day be a valuable and economical energy resource.[118]

The same reasoning leads us to conclude that an intermediate level of difficulty of re-extraction could help put existing separated plutonium in non-weapons-usable form. It could also help convince at least some of the civilian plutonium separating countries to temporarily halt reprocessing until security issues surrounding plutonium can be resolved in a way that greatly reduces the immediate and short-term dangers on as universal a basis as can be achieved.

It may be necessary to offer all countries that own civilian plutonium, but especially Russia and India, a guarantee that grants for plutonium re-extraction would be available should the need arise for using plutonium as an energy source and should it become economical relative to

---

118 NAS 1994, pp. 196–199.

uranium use. Measures to discourage such extraction would also be built into such financial arrangements by holding some of the LEU to be produced by blending down HEU from dismantled weapons as a reserve for use in reactors that would otherwise be fueled with plutonium or with MOX fuel. This LEU reserve could play a global role similar to the domestic role served by the U.S. Strategic Petroleum Reserve. The LEU reserve could be held in part nationally, in part bilaterally (U.S.-Russia), and in part multi-laterally.

We recognize that the proliferation resistance properties of plutonium vitrified with fission products are in some respects stronger than those of plutonium vitrified with actinides or rare earths. We are not advocating that the spent fuel standard be abandoned as an objective. Rather, its refinement so as to accommodate broader goals of putting all weapons-usable plutonium into non-weapons-usable forms is needed. To this end, we strongly urge that vitrification of plutonium with rare earths and actinides should be investigated and that pilot plants should be built.

Reducing the difficulty of re-extraction need not mean lowering the barriers to theft. The difficulty of theft depends on a number of factors, of which a high external gamma radiation field is one of the most important. There are a number of ways in which high external gamma radiation fields can be created to deter theft without mixing plutonium with fission products. One way would be to put plutonium vitrified with rare earths into cesium-137-laced radioactive containers that are manufactured separately. Alternatively, a small container with cesium-137 or a mix of calcined fission products in it could be placed in the canister at a time after plutonium-laden glass has been poured into it. There are some advantages to the former approach. First, the work with gamma emitting fission products can be done entirely separately from the vitrification plant. Second, the canister containing vitrified plutonium can be sealed shortly after the glass is poured. Third, the difficulty of re-extraction may be lower once the glass is removed from the canister. Therefore, this would be more attractive to plutonium-owning countries that regard plutonium as an energy asset, but resistance to theft would still be as high as with the spent fuel standard.

Combining the canister rather than the glass with one or more fission products means that hot-cell processing of gamma-emitting radioactive materials can be done more slowly or even separately from plutonium vitrification. Thereby, achieving the spent fuel standard is made compatible with putting plutonium into a non-weapons-usable form as rapidly as possible. Further, the amount of fission products to achieve a specified gamma radiation field will be far lower, as we have discussed in Chapter 4.

Vitrifying plutonium with a rare earth or actinide and putting gamma-emitting radioactive materials in the canister appears to be the option that best combines various disposition goals. We recommend that DOE commission a feasibility study and appropriate laboratory work on this option in parallel with the pilot plants mentioned above.

Highly Enriched Uranium Disposition Policy

There are about 2,300 metric tons of HEU in the world today, almost all of it in the United States and the former Soviet Union. As we have discussed, about 1,000 metric tons or more of this could become an official surplus as existing arms reduction agreements are implemented over the next decade. It would appear at first that the blending down of HEU to LEU for use in civilian reactors would be the most straightforward way to reduce the attendant security risks. However, as we have noted, the capacity for this blending down does not yet exist in the United States.

The U.S.-Russian agreement, signed in early 1993, will be implemented very slowly, even after blending down actually begins. In each of the first five years, only 10 metric tons of HEU of 90 percent enrichment or greater are required to be blended down, with the rate going up to 30 metric tons per year in the fifteen years after that. At these rates, only 200 metric tons of HEU would have been blended down a decade after the implementation begins. The entire amount will have been blended down in 20 years.

The security threats arising from potential black market sales of HEU may be greater than those arising from plutonium because HEU can be fashioned into weapons of both implosion and "gun-type" designs, while plutonium warheads must be made with an implosion design. Therefore, bilateral or multilateral control, verification of stocks, and adequate materials accounting are all needed.

The large delays in converting HEU into LEU could be mitigated by two policy responses, in addition to the storage and verification arrangements that are needed in any case. The first would be to build a new capacity for blending down HEU as rapidly as compatible with environmental and health considerations. The second would be to vitrify a portion of the HEU. The objective of vitrification would be to quickly raise a barrier to proliferation while leaving open the possibility of recover-ing the HEU and blending it down into LEU for use as fuel. It may be possible to accomplish vitrification on a faster time-scale than blending down might allow. However, vitrification could make LEU derived from HEU uneconomical relative to LEU from newly mined uranium. This is an issue that needs to be addressed prior to a decision on a policy for vitrifying HEU.

Finally, it is relevant to note in this context that HEU does not entirely cease to be a security problem once it has been blended down into LEU. This conversion only raises a barrier to proliferation. Specifically, LEU can be re-enriched to HEU and used for nuclear weapons. If the enrichment process involves the use of gas centrifuge plants, which are in commercial use both in western Europe and in Russia, the detection of re-enrichment at these and any similar plants built in the future will be difficult without extensive new safeguards. It is therefore essential that verification of LEU stocks and, more importantly, of enrichment facilities not now under IAEA safeguards be established so as to make its re-enrichment very difficult. This adds to the need to examine vitrification of HEU as at least a partial disposition option.

In sum, while there is a theoretical solution to the problem of surplus HEU in blending it down to LEU, the practical situation is such that the security threat from HEU will persist, even if we restrict attention to the partial stocks that may be declared surplus over the next decade. Of course, the actual magnitude of the threat is larger, and covers the whole amount (as it does with plutonium).

The question of what should be done with the LEU blended down from HEU is also not as straightforward as might first appear. First, there are commercial pressures to keep the LEU in reserve so as to protect the financial interests of existing commercial producers of uranium as well providers of enrichment services.

Commercial considerations should not be a prime component of the decision to withhold the LEU from the market. However, they could be partly compatible with security criteria. As we have discussed, a portion of the LEU produced from HEU could be used to build up a stock of nuclear power plant fuel. This could be a guarantee to those countries that stop reprocessing spent fuel that they will not lack for fuel, should uranium prices escalate in a manner that is not now anticipated. The strategic stock of LEU could also be used as a modest lever to hold uranium prices to levels that would discourage commercial reprocessing. The first batches of LEU produced by blending down HEU could be devoted to creating such a strategic LEU stock. Any vitrified HEU would also, in effect, serve as a strategic stock. The issue of the size of the stock of LEU required for an effective strategic reserve needs to be studied.

Institutional Issues

It is now at least routinely acknowledged that operations in the nuclear weapons complex must be carried out in conformity with environmental , safety, and health laws and regulations, and with the full participation of the affected communities and other "stakeholders" such as workers. There have also been real and positive changes on a number of other fronts in the Department of Energy, notably at the national level. There are increased opportunities for public participation. Site Specific Advisory Boards are being created or have been created at most nuclear weapons plant sites. Secretary of Energy Hazel O'Leary has released large amounts of data and documents in an unprecedented openness initiative, despite some opposition from the Pentagon. The DOE has also taken the lead in ending funding of the ALMR, a reactor that could produce plutonium, and hence pose a problem from a proliferation standpoint.

This real progress has not yet gone far enough, however. Field and contractor operations and decision-making are not carried out with the openness that is needed; nor is there a sufficient, routine concern for the protection of health and the environment. Spending on weapons continues to be very high, though no weapons are being made. Nuclear weapons testing is re-appearing in new, small-scale disguises. Old technologies that were designated for nuclear weapons production or nuclear power development suddenly appear as clean-up or disposition technologies or both. Most recently, pyroprocessing technology, detached from the now defunct Advanced Liquid Metal Reactor, has made an appearance on the plutonium disposition scene in a forceful manner. The DOE has recently "reprogrammed" money to increase funding of pyroprocessing.[119] While this is ostensibly for examining this technology for plutonium disposition, it also sustains the crucial research aspects of pyroprocessing as a reprocessing complement to the ALMR, which is an advanced plutonium breeder reactor. This is counterproductive for non-prolifera-

---

119 Letter from Joseph Vivona, DOE Chief Financial Officer to Congressman Tom Bevill, Chairman, House Subcommittee on Energy and Water Development, Committee on Appropriations, September 23, 1994.

Policy Issues

tion goals. In short, the hold of nuclear weapons makers and contractors on government policy is still strong.

We have already noted an example of the violation of storage regulations for HEU at Oak Ridge. Incineration is still the basic method of handling mixed radioactive and non-radioactive hazardous wastes. The classification of radioactive waste is still based on a scheme that is not systematically related to the longevity and hazard of the waste. The DOE has yet to submit itself to independent regulation, despite some progress towards creating a framework to achieve this goal. The durability of the progress that has been made is also an open question. In sum, it is not clear that the DOE (the agency whose main mission it was to build bombs), is well-suited to dismantle them and manage the materials. That was the central thrust of the analysis of institutional problems made by the OTA in its 1993 report, which concluded that "U.S. dismantlement and materials management efforts have lacked focus, direction, and coordination."[120] The OTA also concluded that a new office within the DOE or an entirely new agency of government might be needed to manage the problems of the post-Cold War era arising from nuclear weapons dismantlement and materials management.[121]

A new Office of Fissile Materials Disposition was created in January 1994. It is dedicated to disposition issues; it is too early to tell whether this office will be able inspire the kind of work that its mission requires. Further, there is no agreement between the DOE and the Pentagon on crucial disposition issues, such as whether and how much plutonium should be declared a liability, and on how open the government should be with the people of the United States.

It is clear that the vital need to put weapons-usable nuclear materials into non-weapons-usable forms cannot be successfully met, much less with the speed that is desirable from a security standpoint, until these basic institutional issues are resolved. Continued pressure from the affected communities will be central to their resolution.

Public Participation

Successful implementation of plutonium and HEU disposition policy will need the full involvement of the affected communities, especially since speed of implementation is a basic security need. The poor record

120 OTA 1993, p. 122.
121 OTA 1993, p. 13.

on health and environmental protection of the DOE and its predecessor agencies has engendered a profound public distrust that has only begun to be remedied by the openness initiatives of recent years. It is more than likely that any closed process will lead to delays and perhaps also to inappropriate choices of technology. The discussion below is specific to the process for building pilot plants for plutonium vitrification, but the spirit of the comments regarding openness and public participation applies equally well to HEU disposition.

A principal recommendation of this book is that three or four pilot plants for plutonium vitrification should be built. One reason for the emphasis on pilot plants is that the DOE needs to gain operational experience with the technology in order to prepare a sound environmental impact statement on vitrification that would result in the selection of the best way(s) to vitrify various forms of plutonium and the best way to achieve non-proliferation goals in a manner compatible with health and environmental protection.

The pilot plants for vitrifying plutonium should be small enough that they will allow operational experience to be accumulated without risk of severe accidents, but large enough that full-scale plants could be built and operated with confidence based on that experience. In our view, plants meeting these criteria would be large enough that a very open process for setting them up is necessary. On the other hand, it should be possible to design them so that they are small enough to obviate the need for formal environmental impact statements or environmental assessments under the National Environmental Policy Act.

The DOE needs to open up the process for selecting vendors for vitrification technologies so as to include its own laboratories, U.S. corporations that have not been traditional DOE vendors, as well as foreign corporations that have relevant expertise and experience. Further, as we have noted, the DOE should seek to involve Russian institutions at the pilot plant stage so that full-scale plants can be rapidly built in Russia once there is agreement on a disposition method for Russian plutonium. Moreover, the Russian nuclear establishment has considerable operational experience with vitrification and this may help in designing and building the pilot plants.

A closed process of vendor selection would be highly undesirable. The DOE's record in successfully opening major new facilities has been very poor since its creation as a cabinet-level department in 1977. An open process for setting goals for the project, the criteria for vendor selection, and the selection of the vendors is necessary for the critical work of developing plutonium vitrification technology and implementing it in a manner than joins deliberate speed with measures for achieving the protection of health and the environment.

One approach to building plutonium vitrification pilot plants would be for the DOE to hold a competition seeking bids for designing, building, and operating them. The DOE would specify the goals of the project —namely to test and assess for the environmental, safety, health, and economic standpoints of various ways to vitrify plutonium—and the criteria by which these goals would be evaluated. Bids that include qualified Russian collaborators would be encouraged. The proposals would be required to include an environmental and health analysis of the impact of the pilot plants, the employment levels expected, the skills needed, and a summary of the bidders' health and environmental track record. All bids would be made public after the closing date and open discussions would be held at the proposed pilot plant sites on their relative merits. This would aid in the selection of the best proposals. Further, by involving the affected communities, the DOE can ensure their support for the full-scale plutonium vitrification when it is carried out. We suggest that the competition be started in early 1995 and that the selection process of the pilot plant vendors be completed by the end of 1995 or early 1996.

Long-Term Policy Issues

Whether plutonium is vitrified or burned in reactors without reprocessing, a large amount of it will remain for tens of thousands of years. Further, there are large amounts of plutonium in civilian spent fuel as well as in separated plutonium from such spent fuel. These sources of plutonium will constitute a threat to the security of future generations that will endure for thousands of years. As we have discussed, the threat from spent fuel will increase, since the decay of intense gamma emitting isotopes, especially cesium-137, in a few hundred years will make it easier to recover plutonium from spent fuel or glass logs containing high-level waste.

It has generally been assumed that this threat will be greatly reduced by disposing of spent fuel and vitrified waste in a geologic repository. This would increase the costs of plutonium recovery so greatly that it would be more costly to recover spent fuel or vitrified waste from a repository and reprocess it than to derive plutonium from new reactors and reprocessing plants. That should, in fact, be one of the design objectives of a repository.[122]

This theoretical scheme is flawed by one crucial reality. No country has as yet been able to successfully site a geologic repository, though several have been pursuing such a course for decades. There are many reasons for the delays and failures.[123] One principal issue has been that the search for a repository has been bound up in a conflict of interest. The very institutions that have a financial and military interest in nuclear power and nuclear weapons (which together generate most of these wastes) are responsible for or intimately involved in repository selection. Further, we have no institutional experience and not enough scientific knowledge (by a long shot) to predict with confidence the environmental threats that such disposal will pose over hundreds of millennia. The public is rightly skeptical.

If separated plutonium is managed in the interim according to the spent fuel standard, the long-term issues for its disposal are essentially the same as those that arise for unreprocessed spent fuel from civilian power plants. We have already briefly discussed issues of plutonium disposition as they relate to possible repository disposal in the U.S. in Chapter 6.

Essentially complete elimination of plutonium can only be accomplished by two methods. One is to simply wait until the natural radioactive decay of its nuclei have converted it to uranium-235. Since the half-life of plutonium is over 24,000 years, this period of waiting is far longer than the longevity of any human institutions. The other approach is to transmute plutonium using some technique to bombard its nuclei and split them into fission products. Most of these fission products are radioactive; most have half-lives of a few decades or less, but some like technetium-99 and cesium-135, have half-lives that are very long.

So far, the approaches that have been considered for complete transmutation of plutonium in major recent studies have considered only reactor options with some associated reprocessing technology. The two most commonly considered technologies in this category are the Advanced

---

122 It should be noted that plutonium-239 decays into another radioactive material, uranium-235. However, uranium-235 is about 30,000 times less radioactive per unit of weight than plutonium-239, and the radioactivity per canister would be correspondingly smaller. Uranium-235, like plutonium-239, is a weapons-usable fissile material. Therefore, even the decay of plutonium will not end the security threat. The weight of the uranium-235 would be only about 2 percent less than the initial weight of plutonium.

123 For an analysis of the U.S. radioactive waste disposal program, see Makhijani and Saleska 1992.

Liquid Metal Reactor (ALMR), which can also be used to breed plutonium, and a proton accelerator combined with a sub-critical reactor and reprocessing, proposed by Los Alamos National Laboratory. Both technologies must be rejected on proliferation grounds. The U.S. Congress, at DOE's request, has eliminated funding for the ALMR for 1995, though not for the reprocessing technology, called pyroprocessing, associated with it.

One approach that can be used for separated plutonium that does not involve the use of nuclear reactors or reprocessing, but may still result in the elimination of plutonium, is fission using gamma rays. Its feasibility for plutonium disposition has not yet been examined, so far as we have been able to determine. The method involves the fission of plutonium nuclei by the use of high energy gamma rays, which consist of high-energy electromagnetic radiation. The process is called "photofission" because the fission is induced by photons, which are quanta of electromagnetic energy. Other heavy nuclei can also be split by photofission.

A specific spectrum of gamma rays with photons in the energy range of 10 to 15 MeV has a particularly high chance of producing fission in heavy nuclei. This spectrum is called the "giant resonance region" for inducing photofission. Photons of these energies can be produced using an electron accelerator, which is a very well understood technology. The radiation from the stopping (or braking) of high energy electrons (called "bremstrahlung radiation") can be tailored to produce photons in approximately the required spectrum. The photons would induce fission in a plutonium target.

The heat from the braking of the electrons as well as from photofission would have to be carried away by a coolant. This creates the possibility that some of it could be recovered in order to generate electricity. Whether such heat recovery for electricity production is desirable is one of the many questions to be addressed by a feasibility study examining photofission as a long-term disposition option for already separated plutonium.

While the physics of such a scheme is understood, it would be an immensely difficult and complex engineering challenge. During the 1970s, the method was briefly considered for dealing with spent fuel. However, it was rejected because the energy needed to induce photofission to get rid of the long-lived heavy elements would be greater than the energy produced from the fuel in the nuclear reactor.[124] Further, photofission

---

124 Schneider et al., 1974, Vol. 4, Section 9.

would require that the elements to be fissioned be separated from spent fuel, that is, it would require reprocessing. Thus, the approach also is unacceptable for dealing with spent fuel on proliferation grounds.

However, if plutonium is not mixed with fission products, then it can, in principle, be made into targets that are suitable for photofission. The energy use as well as the capital and operating costs of fissioning plutonium completely in this way are likely to be very high. Since the plutonium would be fissioned, the problems of disposing of highly radioactive fission products would also exist with photofission as with all others that depend on fission for plutonium transmutation. The interaction of the photons with a mixture of fission products also needs to be investigated. Further, fission produces neutrons; these neutrons would produce activation products, rendering radioactive a portion of the structure of the devices needed for transmutation. Thus, photofission does not represent a solution to the plutonium disposition problem in the sense of promising something satisfactory without serious long-term financial and environmental costs.

Photofission, if feasible, may offer the potential for complete transmutation of already separated plutonium. But there are many technical unknowns. For instance, its technical feasibility without resorting to some form of reprocessing technology will likely depend on whether appropriate targets can be fabricated that would hold up to the intense radiation and heat until essentially all the plutonium has been fissioned. It is unclear at present whether this can be done in practice.

The only other approach that could get rid of separated plutonium without reprocessing is to shoot it into the sun. While at present both costs and dangers of this approach are immense, we believe this also deserves a more careful feasibility study.

Neither space disposal nor photofission can deal with the problem of plutonium in spent fuel, unless it is first reprocessed. Therefore, when examined from the perspective of the overall problem of plutonium elimination, they do not represent solutions. Whether such technologies would be worthwhile at all just for disposing of already separated plutonium is an open question. We believe that both approaches deserve serious feasibility studies so that we may have a basis to decide whether some research and development of one or both of them would be worthwhile.

The future of security and environmental issues arising from the creation of plutonium is bound up with nuclear power production, since essentially all nuclear power plants produce large quantities of plutonium as a normal part of their operation. The only exceptions to this are reactors that use HEU as fuel, but this fuel is itself a proliferation problem. Therefore, if we are to make an attempt to definitively deal with the threats arising from the existence of weapons-usable fissile materials, we must confront the central issue of what energy sources the world will rely on for the long-term. Our final recommendation is, therefore, that the use of nuclear power should be more carefully evaluated in light of the long-term proliferation problems posed by the very existence of large and increasing quantities of plutonium in spent fuel.

References

ABB 1993 — ABB Combustion Engineering, *DOE Plutonium Disposition Study: Pu Consumption in ALWRs*, DOE/SF/19682, Windsor, Conn., May 15, 1993.

Albright et al. 1993 — David Albright, Frans Berkhout, and William Walker, *World Inventory of Plutonium and Highly Enriched Uranium 1992*, Oxford University Press, 1993.

Bates et al. 1990 — J.K. Bates, W.H. Ebert, and T.J. Gerding, "Vapor hydration and subsequent leaching of transuranic-containing SRL and WV Glasses," in *The Proceedings of the Conference on High-Level Radioactive Waste Management*, held at Las Vegas, April 8–12, 1990, American Nuclear Society, La Grange Park, Illinois, 1990.

Bates 1992 — J.K. Bates, "Colloid formation during waste form reaction: Implications for nuclear waste disposal," *Science*, May 1, 1992.

Benedict et al. 1981 — Manson Benedict, Thomas H. Pigford, and Hans Wolfgang Levi, *Nuclear Chemical Engineering, Second Edition*, McGraw-Hill, New York, 1981.

Berkhout et al. 1992 — Frans Berkhout, Anatoli Diakov, Harold Feiveson, Helen Hunt, Marvin Miller, and Frank von Hippel, "Disposition of separated plutonium." Center for Energy and Environmental Studies, PU/CEES Report Number 272, Princeton University, Princeton, NJ, September 1992.

Bibler 1982 — N.E. Bibler, "Effects of alpha, gamma, and alpha-recoil radiation on borosilicate glass containing Savannah River Plant defense high-level nuclear waste," in *Scientific Basis for Nuclear Waste Management*, S.V. Topp (ed.), Elsevier, New York, 1982, pp. 681–87.

References

Bibler and Howitt 1988	N.E. Bibler and D.G. Howitt. "Radiation effects in silicate glasses—A review," in *Materials Stability and Environmental Degradation*, A. Barkatt, et al. (ed.), Materials Research Society, Pittsburgh, Pennsylvania, 1988, pp. 263–284.
Bickford 1990	Dennis F. Bickford. "Advanced radioactive glass melters (U)." Westinghouse Savannah River Co., Aiken, South Carolina, April 22, 1990.
Brandon 1993	Norman E. Brandon from Nuclear Fuel Services, "Enrichment Blending: An Overview and Analysis," a paper presented to the US Council for Energy Awareness International Enrichment Conference, June 14, 1993.
Chow and Solomon 1993	Brian G. Chow and Kenneth A. Solomon, *Limiting the Spread of Weapon-Usable Fissile Materials*, National Defense Research Institute, RAND, Santa Monica, CA, 1993.
DNFSB 1994	Defense Nuclear Facilities Safety Board, *Plutonium Storage Safety at Major Department of Energy Facilities*, Defense Nuclear Facilities Safety Board Technical Report Number DNFSB/TECH-1, Washington, D.C., April 14, 1994.
DOE 1994	Department of Energy, *Openness Press Conference Fact Sheets*, Washington, D.C. June 27, 1994.
DOE 1994a	Department of Energy, *Plutonium Working Group Report on Environmental, Safety and Health Vulnerabilities Associated with the Department's Plutonium Storage*, Draft, Publication Number DOE/EH-0415, U.S. Department of Energy, Washington, D.C., September 1994.
Gibbons 1994	John H. Gibbons, Testimony before the Committee on Energy and Natural Resources, U.S. Senate, May 26, 1994.
ERF and NRDC 1992	Energy Research Foundation and Natural Resources Defense Council, *Rethinking Plutonium: A Review of Plutonium Operations in the U.S. Nuclear Weapons Complex*, Energy Research Foundation, Columbia, South Carolina, April 1992.

Forsberg et al. 1994	C.W. Forsberg, E.C. Beahm, and G.W. Parker, "Direct Conversion of Radioactive and Chemical Waste Containing Metals, Ceramics, Amorphous Solids, and Organics to Glass," paper presented to the Spectrum '94 Nuclear and Hazardous Waste Management International Topical Meeting, Atlanta, Georgia, May 16, 1994. Copies may be obtained from the principal author at Oak Ridge National Laboratory, Oak Ridge, TN.
GAO 1992	General Accounting Office, *Nuclear Waste: Defense Waste Processing Facility—Cost, Schedule, and Technical Issues*, General Accounting Office, GAO/RCED-92-183, Washington, D.C., June 1992.
Grumbly 1994	Thomas P. Grumbly, Written remarks presented at the National Symposium on Weapons-Usable Fissile Materials in Washington, D.C., sponsored by the Institute for Energy and Environmental Research, November 17, 1994.
Hersh 1991	Seymour M. Hersh, *The Samson Option: Israel's Nuclear Arsenal and American Foreign Policy*, Random House, New York, 1991.
IEER 1994	Summary of Presentations at the National Symposium on Weapons-Usable Fissile Materials on November 17 and 8, 1994 held in Washington, D.C., Institute for Energy and Environmental Research, Takoma Park, Maryland, January 1995.
IPPNW and IEER 1992	International Physicians for the Prevention of Nuclear War and the Institute for Energy and Environmental Research, *Plutonium: Deadly Gold of the Nuclear Age*, International Physicians Press, Cambridge, MA, 1992.
LaMarsh 1983	John R. LaMarsh, *Introduction to Nuclear Engineering*, Second Edition, Addison-Wesley, Reading, MA, 1983.
Leventhal and Dolley 1994	Paul Leventhal and Steven Dolley, *A Japanese Strategic Uranium Reserve: A Safe and Economic Alternative to Plutonium*, Nuclear Control Institute, Washington, D.C., January 14, 1994.
Makhijani and Tucker 1985	Arjun Makhijani and Kathleen Tucker, *Heat, High Water and Rock Instability at Hanford: A Preliminary Assessment of the Suitability of the Hanford,*

	Washington Site for a High-Level Nuclear Waste Repository, Health and Energy Institute, Washington, D.C. 1985. Copies available from the Institute for Energy and Environmental Research, Takoma Park, MD.
Makhijani 1991	Arjun Makhijani, *Glass in the Rocks: Some Issues Concerning the Disposal of Radioactive Borosilicate Glass in a Yucca Mountain Repository*, Institute for Energy and Environmental Research, Takoma Park, MD, January 29, 1991.
Makhijani and Saleska 1992	Arjun Makhijani and Scott Saleska, *High-level Dollars, Low-level Sense: A Critique of Present Policy for Management of Long-lived Radioactive Waste and Discussion of an Alternative Approach*, Apex Press, New York, 1992
Moncouyoux et al 1991	J.P. Moncouyoux, R. Boen, M. Puyou, and A. Jouan, "New Vitrification Technologies, Rhone Valley Research Center VALHRO), 30205 Bagnols-sur-Cèze Cedex, France, 1991.
NAS 1994	National Academy of Sciences, *Management and Disposition of Excess Weapons Plutonium*, Committee on International Security and Arms Control, Washington, D.C., 1994.
NASA 1982	National Aeronautics and Space Administration, *Analysis of Space Systems for the Space Disposal of Nuclear Waste: Follow-on Study*, NASA CR161992, NASA/Marshall Space Flight Center, Huntsville, Alabama, 1982.
National Research Council 1983	National Academy of Sciences National Research Council, *A Study of the Isolation System for Geologic Disposal of Radioactive Waste*, Waste Isolation Systems Panel, Washington, D.C., 1983
ORNL 1992	Oak Ridge National Laboratory, *Characteristics of Potential Repository Wastes*, Office of Civilian Radioactive Waste Management, DOE/RW-0184-R1, 3 volumes, July 1992.
OTA 1986	Office of Technology Assessment, *Staff paper on the Sub-seabed disposal of high-level waste*, U.S. Government Printing Office, Washington, D.C., 1986.

OTA 1991	Office of Technology Assessment, *Long-lived Legacy: Managing High-Level and Transuranic Waste at the DOE Weapons Complex*, background paper, U.S. Government Printing Office, Washington, D.C., May 1991.
OTA 1993	Office of Technology Assessment, *Dismantling the Bomb and Managing the Materials*. Washington, D.C.: U.S. Government Printing Office, September 1993.
Plodinec 1979	M.J. Plodinec, *Development of Glass Compositions for Immobilization of SRP Waste*, U. S. Department of Energy report DP-1517, Savannah River Laboratory, Aiken, South Carolina, 1979.
Schneider et al. 1974	K.J. Schneider et al. *High-Level Radioactive Waste Management Alternatives*, 4 volumes, Battelle Pacific Northwest Laboratories, Richland, WA, 1974.
Sylvester 1994	Kory William Budlong Sylvester, *A Strategy for Weapons-grade Plutonium Disposition*, thesis prepared for an M.S. in Nuclear Engineering and an M.S. in Technology and Policy, Massachusetts Institute of Technology, Cambridge, MA, September 1994.
UEO 1990	Uranium Enrichment Organization, *The Ultimate Disposition of Depleted Uranium*, Oak Ridge, TN, 1990.
Von Hippel et al. 1993	Frank von Hippel, Marvin Miller, Harold Feiveson, Anatoli Diakov, and Frans Berkhout, "Eliminating Nuclear Warheads," *Scientific American*, August 1993, pp. 44-49.
Westinghouse 1993	Vitrification of Excess Plutonium, Plutonium Vitrification Task Group, WSRC-RP-93-755, Westinghouse Savannah River Company, Aiken, SC, May 1993, p. 25.
Wronkiewicz 1994	D.J. Wronkiewicz, "Radionuclide Decay Effects on Waste Glass Corrosion," in *Materials Research Society Symposium Procedures*, Vol. 333. Materials Research Society.
WVDP 1990	West Valley Demonstration Project, *Project Overview*, presented to P. Johnson and R. Morgan, representing the Office of Technology Assessment, West Valley, New York, February 21, 1990.

Glossary

Alpha decay: The emission of a nucleus of a helium atom from the nucleus of an element, generally of a heavy element, in the process of its radioactive decay.

Alpha particle: The nuclei of a helium atom (with two neutrons and two protons each) that are discharged by radioactive decay of many heavy elements, such as uranium-238 and plutonium-239.

Atomic number (symbolized Z): The number of protons in a nucleus. It determines the chemical properties of an element.

Beta decay: The emission of electrons or positrons (particles identical to electrons, but with a positive electrical charge) from the nucleus of an element in the process of radioactive decay of the element.

Beta particle: Electrons or positrons (positively charged electrons) emitted by many radionuclides in the course of radioactive decay.

Curie: Unit of radioactivity equal to the radioactivity of 1 gram of radium-226. It is equal to 37 billion disintegrations per second.

Decommissioning: Decontamination and dismantlement of retired, contaminated facilities and removal and/or disposal of the resulting wastes.

DWPF: Defense Waste Processing Facility, the name of the vitrification plant for high-level radioactive wastes at the Savannah River Site.

Fissile Material: A material consisting of atoms whose nuclei can be split when irradiated with low energy (ideally, zero energy) neutrons.

Gamma radiation: Electromagnetic waves released during radioactive decay that can ionize atoms and split chemical bonds. Gamma rays are similar to X-rays, the latter term being applied usually to electromagnetic waves generated by electron accelerators, as for instance in medical equipment.

HEU: Highly enriched uranium.

Isotope: Atoms of the same element that have the same number of protons (and hence the same chemical properties), but a different number of neutrons, and therefore, different atomic weights.

LEU: Low enriched uranium.

Low-level radioactive waste: A catch-all category of waste defined by U.S. law as all wastes that are not in other categories such as "high-level" waste and mill tailings; radioactivity of "low-level" waste varies widely and includes both short- and long-lived isotopes.

Mass number (symbolized A): The sum of the number of protons and the number of neutrons in a nucleus.

Mill tailings: A slurry of about 40 percent solids (including radioactive particles and chemically hazardous metals) and 60 percent liquid, primarily water.

Metric ton: 1,000 kilograms or about 2,204 pounds. The usual U.S. ton measurement, called a short ton, is 2,000 pounds.

MOX: A fuel composed of a mixture of plutonium dioxide and uranium dioxide.

NPT: The nuclear Non-Proliferation Treaty.

Reprocessing: The chemical separation of irradiated nuclear fuel into uranium, plutonium, and fission products.

Specific activity: A measure of the radioactivity of a unit weight (generally one gram) of material.

Spontaneous fission: The spontaneous splitting of the nucleus into two new nuclei, generally with the emission of one or more neutrons and the release of energy.

WVDP: West Valley Demonstration Plant, the name of the vitrification plant for high-level radioactive wastes at West Valley, New York.

Appendix A

Text of a Letter on Plutonium Sent to President Clinton on October 19, 1994

October 19, 1994

President Clinton
The White House
Washington, D.C.

Dear Mr. President,

The growing global stockpile of surplus military and civilian plutonium presents one of the most serious security problems that we face. In particular, how we handle the surplus plutonium problem today will affect the shape of the world for generations to come. It is imperative that the United States seize the moment to provide the leadership that it will take to definitively end the plutonium threat.

We recognize the important steps that your administration has taken to remedy past neglect of this issue, but we feel that it is extremely important to make it a far more urgent priority and to adopt a more comprehensive policy. *The first step towards such a policy is to declare all excess military and civilian plutonium a security, economic, and environmental liability.*

Nature of the Threat

Knowledge of nuclear weapons technology is now so widespread that it is not a substantial barrier to nuclear proliferation. Rather, as the January 1994 National Academy of Sciences report on plutonium noted, limiting "access to fissile materials is the principal technical barrier to proliferation in today's world . . . ." All separated plutonium, whether of military or civilian origin, can be used to make nuclear weapons. Therein lies the central aspect of the security threat from plutonium.

Plutonium is also an environmental threat because it is a highly carcinogenic, radioactive substance, with a half-life of over 24,000 years. As such, it can be used to make radiation-dispersion weapons. These weapons require only conventional explosives to disperse the plutonium over a wide area. A report to the U.S. Joint Chiefs of Staff written in 1947, said that such contamination would be a very effective means of terror because it would stimulate "man's primordial fears, those of the unknown, the invisible, the mysterious."

Mr. President, as you know only too well, the control, safe storage, and disposal of plutonium take on special urgency in view of the severe and frequent political, military, and economic crises in the former Soviet Union. The danger signs of a potential black market in fissile materials have been evident for some time. For example, there have been many arrests in Germany of people attempting to smuggle radioactive materials originating in the former Soviet Union. A small quantity of weapon-grade plutonium of Russian origin was recently confiscated in Germany; it may have been a sample of a larger amount being offered for clandestine sale. That is the most ominous sign so far of a clandestine traffic in fissile materials.

Russia continues to separate plutonium from civilian reactor spent fuel at its Chelyabinsk-65 plant in the Urals. Plutonium oxide, the chemical form in which reprocessing plants generally produce plutonium, is particularly suited to radiation-dispersal bombs. Some 30 tons of separated plutonium sit near Chelyabinsk—a constant temptation to illegal sales in the context of a deteriorating economy. The amount is growing due to continued production from civilian reactors, which is not covered by the recent U.S.-Russian agreement to stop military plutonium production at Tomsk-7. Organized crime appears to have penetrated the conventional arms market in the East. A fusion of the two trends—conventional arms traffic and traffic in radioactive materials—could be disastrous.

Plutonium is also an economic liability, even though it can, in principle, be used as an energy source. In practice, however, plutonium is a very expensive nuclear fuel because it is highly radioactive, requiring special handling to process, in comparison to uranium. Moreover, inexpensive uranium resources have turned out to be abundant, and will remain so for the foreseeable future. Authoritative analyses, notably the January 1994 report by the National Academy of Sciences, have concluded that even if the plutonium is "free"—that is, even when the sunk costs of producing plutonium are entirely ignored—the cost of processing plutonium for use in reactors is far greater than its market value as a nuclear fuel. Natural uranium prices would have to go up several times in order for plutonium to become an economical fuel. This is highly unlikely, based on considerations of uranium resource availability.

In sum, the disposal of excess plutonium will result in substantial net costs, whether or not it is used as a fuel. This makes it different from highly enriched uranium, which poses a comparable security threat, but which can be diluted and made into low enriched uranium for use as a civilian reactor fuel. Of course, it is essential that environmental and safety rules are carefully observed. The process of dealing with highly enriched uranium should proceed with as high a priority as plutonium, but on a different track because, diluted to a low enrichment, it is potentially an economical energy resource, if used in existing reactors to displace newly mined uranium. Low enriched uranium, while suitable for civilian power reactors, cannot be used to make nuclear weapons without costly, difficult, and detectable re-enrichment.

The United States stopped military plutonium production in 1988 and made that a formal policy in 1991. Well before that the U.S. had wisely abandoned the use of plutonium in civilian reactors as impractically costly and problematic for global security. It has therefore long been the *de facto* position of the U.S. government that plutonium is an economic and security liability. However, the U.S. has failed to gain any foreign policy advantage from this position, because it has not formalized this policy.

There are many governments, notably Russia, France, the U.K., Japan, and India, that take the position that plutonium is a valuable energy resource; they continue to produce it despite the abundance of inexpensive uranium. As the National Academy of Sciences study on plutonium noted, such a view is largely based on the inertia of long-standing breeder reactor and plutonium programs. These programs were designed in the 1950s when it was generally believed that uranium would be a very scarce resource whose use should be maximized by the introduction of plutonium into the civilian economy. The U.S. cannot exercise the full potential of its influence on these countries to adopt a more pragmatic economic and security perspective regarding plutonium until it makes a formal declaration about the status of its own excess plutonium.

A Declaration that Plutonium is a Liability

We request you to issue a formal declaration that all excess military and civilian plutonium is a liability from the security, economic, and environmental points of view. Excess plutonium should be defined as all plutonium, of military or civilian origin, in any chemical form, that is not a physical component of the weapons designated as part of the U.S. arsenal in the Nuclear Stockpile Memorandum. Such a declaration of principle would not prejudge the method that would be used to process the excess plutonium, other than excluding reactors that must be accompanied by technology to separate plutonium from spent fuel as well as plutonium separation technology. A declaration would make U.S. and global security, safe storage and dismantlement, and protection of health and the environment—not energy production—the basis for making a decision on long-term plutonium disposition. A draft declaration is attached for your consideration.

Such a definition of the status of excess plutonium as a liability would mean the following:

Excess plutonium would not be used to make nuclear weapons.

Excess plutonium would not be regarded as part of an energy program. This does not *a priori* preclude using nuclear reactors as part of the disposition decision, but it does exclude any plutonium separation from spent fuel. The choice of a plutonium disposition option would be made on the basis of security and environmental criteria and not on the energy value of plutonium.

The U.S. government would make it a priority to persuade other countries to join in a similar declaration, even if their current position is that plutonium is a valuable energy resource. Cooperative exploration of energy policy issues would be an important part of this plutonium diplomacy. The initiative of the U.S. and other countries for a cut-off of military fissile materials production in the context of the negotiations for the extension of the Non-Proliferation Treaty would provide one suitable immediate context for the discussions.

The world's stockpile of separated plutonium now totals about 400 metric tons. About 40 metric tons or less are required for a world nuclear arsenal of about 10,000 weapons or less. Moreover, the figure of 10,000 weapons is at the higher end of nuclear arsenals advocated by military analysts; many advocate far lower numbers. It is urgent that secure, verifiable storage of excess plutonium be implemented. Plutonium should then be put into a form that makes it impossible to use in nuclear or radiation weapons without costly, difficult, and dangerous processing that could be detected with relative ease.

The U.S. is the only leading country that has nothing to lose and everything to gain by a declaration that plutonium is a liability. The other nuclear weapons powers (except perhaps China) have great bureaucratic vested interests and momentum in preserving their plutonium production programs. An early U.S. declaration would also empower the thousands of non-governmental organizations, parliamentarians, scientists, engineers, economists, physicians, and others around the world who are seeking to end the production of plutonium. This would deepen and consolidate the policies you initiated early in your administration to limit the growth of plutonium production. If we are successful in this endeavor of curbing and ending the threat of plutonium proliferation, as indeed we must be, it will be an enduring contribution of your administration to world peace for generations to come.

Thank you very much for taking the time to consider this letter. If you would like further information or if you have any questions, please contact Dr. Arjun Makhijani at the Institute for Energy and Environmental Research in Takoma Park, Maryland at (301) 270-5500.

We look forward to hearing from you on this vital matter of great importance to all of us and to future generations.

Respectfully Yours,

(The list of organizational and individual co-signers is attached.)

cc: Assistant to the President for Science and Technology, John Gibbons
Assistant to the President for National Security Affairs, Anthony Lake
Secretary of Energy, Hazel O'Leary

Appendix A. Text of a Letter Sent to President Clinton

U.S. Government Declaration: Excess Plutonium is a Liability

The growing global stockpile of surplus military and civilian plutonium presents one of the most serious security problems that we face. How we handle this problem today will affect the shape of the world for generations to come. It is imperative that the United States provide the leadership to definitively end the plutonium threat.

The United States stopped military plutonium production in 1988 and made that a formal policy in 1991. Well before that it had abandoned the use of plutonium in civilian reactors as impractically costly and problematic for global security. It has therefore long been the *de facto* position of the U.S. government that plutonium is an economic and security liability. The main purpose of this declaration is to formalize that policy in order to promote U.S. and global security.

There are many governments that maintain that plutonium is a valuable energy resource; they continue to produce it despite the abundance of inexpensive uranium. As the January 1994 National Academy of Sciences study on plutonium noted, such a view is based, in part, on the inertia of long-standing breeder reactor and plutonium programs. These programs were designed in the 1950s when it was generally believed that uranium would be a very scarce resource. The U.S. cannot realize the full potential of its influence on these countries to adopt a more pragmatic economic and security perspective on plutonium until it makes a formal declaration about the status of its own excess plutonium.

A Declaration that Plutonium is a Liability

It is U.S. government policy, based on extensive evidence and analysis, including the January 1994 study by the National Academy of Sciences, that all excess plutonium is a security, economic, and environmental liability. Excess plutonium is defined as all plutonium, of military or civilian origin, in any chemical form, that is not a physical component of the weapons designated as part of the U.S. arsenal in the Nuclear Stockpile Memorandum. It has the following operational meaning for policy:

Excess plutonium will not be used to make nuclear weapons.

Excess plutonium will not be regarded as part of an energy program. This does not *a priori* preclude using nuclear reactors as part of the disposition decision, but it does exclude any plutonium separation from spent fuel. The choice of a plutonium disposition option would be made on the basis of security and environmental criteria, and not on the energy value of plutonium.

The U.S. government will make it a priority to persuade other countries to join in a similar declaration, even if their current position is that plutonium is a valuable energy resource. Cooperative exploration of energy policy issues will be an important part of U.S. diplomacy on plutonium and highly enriched uranium, the other fissile material of great concern to our security.

Co-Signatories

American Friends Service Committee (Colorado)
Americans for Indian Opportunity
Mavis Belisle, Peace Farm*
Daniel Cantor, The New Party*
Center for Defense Information
Center for International Environmental Law
Citizens Clearinghouse for Hazardous Wastes, Inc.
Concerned Citizens for Nuclear Safety
Earth Action
Energy Research Foundation
Environmental Research Foundation
Friends of the Earth
Government Accountability Project
Greenpeace
Hanford Education Action League (HEAL)
Institute for Defense & Disarmament Studies
International Women's Peace Initiative
Jobs with Peace
Mim Kelber & Bella Abzug, Women's Environmental & Development Organization (WEDO)*
Eliza Klose, Institute for Sov-Amer Relations (ISAR)*
Manhattan Project II
National Committee for Radiation Victims
Native Americans for a Clean Environment (NACE)
Native Youth Alliance
Network of East-West Women
Ken Nichols, Audubon Naturalist Society*
Nuclear Control Institute
Nuclear Information & Resource Service (NIRS)
Nuclear Safety Campaign
Oak Ridge Environmental Peace Alliance
Panhandle Area Neighbors and Landowners (PANAL)
Peace Action
Physicians for Social Responsibility
Plutonium Free Future
Public Citizen
Serious Texans Against Nuclear Dumping (STAND, Inc.)
Ted Taylor
Southwest Research and Information Center
Tri-Valley CAREs
20/20 Vision
Sima Wali, Refugee Women in Development*
Water Information Network
Women's Action for New Directions
Women Strike for Peace

*Organization is listed for identification purposes.

Appendix B
Physical, Nuclear, and Chemical Properties of Plutonium

Plutonium-239 is one of the two fissile materials used for the production of nuclear weapons. The other fissile material is uranium-235. Plutonium-239 is virtually nonexistent in nature. It is made by bombarding uranium-238 with neutrons in a nuclear reactor. Uranium-238 is present in quantity in most reactor fuel; hence plutonium-239 is continuously made in these reactors. Since plutonium-239 can itself be split by neutrons to release energy, plutonium-239 provides a portion of the energy generated in a nuclear reactor.

The physical properties of plutonium metal are summarized in Table 6.

TABLE 6. **Physical characteristics of plutonium metal**

Color:	silver
Melting point:	641° C
Boiling point:	3,232° C
Density:	16 to 20 grams/cubic centimeter

Nuclear Properties of Plutonium

Plutonium belongs to the class of elements called transuranic elements whose **atomic number**[*] is higher than 92, the atomic number of uranium. Essentially all transuranic materials in existence are manmade. The atomic number of plutonium is 94.

Plutonium has 15 isotopes with **mass numbers** ranging from 232 to 246. Isotopes of the same element have the same number of protons in their nuclei but differ by the number of neutrons. Since the chemical characteristics of an element are governed by the number of protons in the nucleus, which equals the number of electrons when the atom is electrically neutral (the usual elemental form at room temperature), all isotopes have nearly the same chemical characteristics. This means that in most cases it is very difficult to separate isotopes from each other by chemical techniques.

[*]Terms in bold are defined in glossary.

Only two plutonium isotopes have commercial and military applications. Plutonium-238, which is made in nuclear reactors from neptunium-237, is used to make compact thermo-electric generators; plutonium-239 is used for nuclear weapons and for energy; plutonium-241, although fissile (see next paragraph), is impractical both as a nuclear fuel and a material for nuclear warheads. Some of the reasons are far higher cost, shorter half-life, and higher radioactivity than plutonium-239. Isotopes of plutonium with mass numbers 240 through 242 are made along with plutonium-239 in nuclear reactors, but they are contaminants with no commercial applications. In this fact sheet we focus on civilian and military plutonium (which are interchangeable in practice—see Table 5), which consist mainly of plutonium-239 mixed with varying amounts of other isotopes, notably plutonium-240, -241, and -242.

Plutonium-239 and plutonium-241 are fissile materials. This means that they can be split by both slow (ideally zero-energy) and fast neutrons into two new nuclei (with the concomitant release of energy) and more neutrons. Each fission of plutonium-239 resulting from a slow neutron absorption results in the production of a little more than two neutrons on the average. If at least one of these neutrons, on average, splits another plutonium nucleus, a sustained chain reaction is achieved.

The even isotopes, plutonium-238, -240, and -242 are not fissile but yet are fissionable—that is, they can only be split by high energy neutrons. Generally, fissionable but non-fissile isotopes cannot sustain chain reactions; plutonium-240 is an exception to that rule.

The minimum amount of material necessary to sustain a chain reaction is called the critical mass. A super-critical mass is bigger than a critical mass, and is capable of achieving a growing chain reaction where the amount of energy released increases with time.

The amount of material necessary to achieve a critical mass depends on the geometry and the density of the material, among other factors. The critical mass of a bare sphere of plutonium-239 metal is about 10 kilograms. It can be considerably lowered in various ways.

The amount of plutonium used in fission weapons is in the 3 to 5 kilograms range. According to a recent Natural Resources Defense Council report,[125] nuclear weapons with a destructive power of 1 kiloton can be built with as little as 1 kilogram of weapon grade plutonium.[126] The smallest theoretical critical mass of plutonium-239 is only a few hundred grams.

In contrast to nuclear weapons, nuclear reactors are designed to release energy in a sustained fashion over a long period of time. This means that the chain reaction must be controlled—that is, the number of neutrons produced needs

---

125 Cochran, Thomas B. and Christopher E. Paine, *The Amount of Plutonium and Highly Enriched Uranium Needed for Pure Fission Nuclear Weapons* Natural Resources Defense Council, Washington, DC, 22 August 1994.

126 For comparison the bomb dropped on Nagasaki on August 9, 1945 had 6.1 kg of plutonium and a destructive power of about 20 kilotons.

Appendix B. Properties of Plutonium

to equal the number of neutrons absorbed. This balance is achieved by ensuring that each fission produces exactly one other fission.

All isotopes of plutonium are radioactive, but they have widely varying half-lives. The half-life is the time it takes for half the atoms of an element to decay. For instance plutonium-239 has a half-life of 24,110 years while plutonium-241 has a half-life of 14.4 years. The various isotopes also have different principal decay modes. The isotopes present in commercial or military plutonium-239 are plutonium-240, -241, and -242. Table 7 shows a summary of the radiological properties of five plutonium isotopes.

TABLE 7. Important plutonium isotopes' radiological properties

	Pu-238	Pu-239	Pu-240	Pu-241	Pu-242
Half-life (in years)	87.74	24,110	6,537	14.4	376,000
Specific activity (curies/gram)	17.3	0.063	0.23	104	0.004
Principal decay mode	alpha	alpha	alpha some spontaneous fission[a]	beta	alpha
Decay energy (MeV)	5.593	5.244	5.255	0.021	4.983
Radiological hazards	alpha, weak gamma	alpha, weak gamma	alpha, weak gamma	beta, weak gamma[b]	alpha weak gamma

Source: CRC Handbook of Chemistry and Physics, 1990-1991. Various sources give slightly different figures for half-lives and energies.

[a] Source of neutrons causing added radiation dose to workers in nuclear facilities. A little spontaneous fission occurs in most plutonium isotopes.

[b] Plutonium-241 decays into americium-241, which is an intense gamma-emitter.

The isotopes of plutonium that are relevant to the nuclear and commercial industries decay by the emission of alpha particles, beta particles, or **spontaneous fission**. **Gamma radiation**, which is penetrating electromagnetic radiation, is often associated with **alpha and beta decays**.

Chemical Properties and Hazards of Plutonium

Table 8 describes the chemical properties of plutonium in air. These properties are important because they affect the safety of storage and of operation during processing of plutonium. The oxidation of plutonium represents a health hazard since the resulting stable compound, plutonium dioxide, is in particulate form that can be easily inhaled. It tends to stay in the lungs for long periods, and is also transported to other parts of the body. Ingestion of plutonium is considerably less dangerous since very little is absorbed while the rest passes through the digestive system.

TABLE 8. How plutonium metal reacts in air

FORMS & AMBIENT CONDITIONS	REACTION
Non-divided metal at room temperature (corrodes)	relatively inert, slowly oxidizes
Divided metal at room temperature (PuO_2)	readily reacts to form plutonium dioxide
Finely divided particles under about 1 millimeter diameter	spontaneously ignites at about 150° C[c]
particles over about 1 millimeter diameter	spontaneously ignites at about 500° C
Humid, elevated temperatures (PuO_2)	readily reacts to form plutonium dioxide

[c] US Department of Energy, "Assessment of Plutonium Storage Safety Issues at Department of Energy Facilities," DOE/DP-0123T (Washington, DC: US DOE, January 1994).

Important Plutonium Compounds and Their Uses

Plutonium combines with oxygen, carbon, and fluorine to form compounds which are used in the nuclear industry, either directly or as intermediates.

Table 10 shows some important plutonium compounds. Plutonium metal is insoluble in nitric acid and plutonium dioxide is slightly soluble in hot, concentrated nitric acid. However, when plutonium dioxide and uranium dioxide form a solid mixture, as in spent fuel from nuclear reactors, then the solubility of plutonium dioxide in nitric acid is enhanced due to the fact that uranium dioxide is soluble in nitric acid. This property is used when **reprocessing** irradiated nuclear fuels.

Appendix B. Properties of Plutonium

TABLE 9. Important plutonium compounds and their uses

COMPOUND	USE
Oxides	
Plutonium Dioxide (PuO_2)	can be mixed with uranium dioxide (UO_2) for use as reactor fuel
Carbides	
Plutonium Carbide (PuC)	all three carbides can potentially be used as fuel in breeder reactors
Plutonium Dicarbide (PuC_2)	
Diplutonium Tricarbide (Pu_2C_3)	
Fluorides	
Plutonium Trifluoride (PuF_3)	both fluorides are intermediate compounds in the production of plutonium metal
Plutonium Tetrafluoride (PuF_4)	
Nitrates	
Plutonium Nitrates [$Pu(NO_3)_4$] and [$Pu(NO_3)_3$]	no use, but it is a product of reprocessing (extraction of plutonium from used nuclear fuel)

Formation and Grades of Plutonium-239

Plutonium-239 is formed in both civilian and military reactors from uranium-238.

The subsequent absorption of a neutron by plutonium-239 results in the formation of plutonium-240. Absorption of another neutron by plutonium-240 yields plutonium-241. The higher isotopes are formed in the same way. Since plutonium-239 is the first in a string of plutonium isotopes created from uranium-238 in a reactor, the longer a sample of uranium-238 is irradiated, the greater the percentage of heavier isotopes. Plutonium must be chemically separated from the fission products and remaining uranium in the irradiated reactor fuel. This chemical separation is called reprocessing.

Fuel in power reactors is irradiated for longer periods at higher power levels, called high "burn-up", because it is fuel irradiation that generates the heat required for power production. If the goal is production of plutonium for military purposes then the "burn- up" is kept low so that the plutonium-239 produced is as pure as possible, that is, the formation of the higher isotopes, particularly plutonium-240, is kept to a minimum.

Plutonium has been classified into grades by the U.S. DOE (Department of Energy) as shown in Table 10.

TABLE 10. Grades of plutonium

GRADE	% Pu-240 CONTENT
Supergrade	2–3 %
Weapon grade	< 7 %
Fuel grade	7–19 %
Reactor grade	19 % or greater

It is important to remember that this classification of plutonium according to grades is somewhat arbitrary. For example, although "fuel grade" and "reactor grade" are less suitable as weapons material than "weapon grade" plutonium, they can also be made into a nuclear weapon, although the yields are less predictable because of unwanted neutrons from spontaneous fission. The ability of countries to build nuclear arsenals from reactor grade plutonium is not just a theoretical construct. It it is a proven fact. During a June 27, 1994 press conference, Secretary of Energy Hazel O' Leary revealed that in 1962 the United States conducted a successful test with "reactor grade" plutonium. All grades of plutonium can be used as weapons of radiological warfare which involve weapons that disperse radioactivity without a nuclear explosion.

Appendix C

Uranium: Its Uses and Hazards

First discovered in the 18th century, uranium is an element found everywhere on Earth, but mainly in trace quantities. In 1938, German physicists Otto Hahn and Fritz Strassmann showed that uranium could be split into parts to yield energy. Uranium is the principal fuel for nuclear reactors and the main raw material for nuclear weapons.

Natural uranium consists of three **isotopes**[*]: uranium-238, uranium-235, and uranium-234. Uranium isotopes are radioactive. The nuclei of radioactive elements are unstable, meaning they are transformed into other elements, typically by emitting particles (and sometimes by absorbing particles). This process, known as radioactive decay, generally results in the emission of **alpha** or **beta particles** from the nucleus. It is often also accompanied by emission of **gamma radiation**, which is electromagnetic radiation, like X-rays. These three kinds of radiation have very different properties in some respects but are all ionizing radiation—each is energetic enough to break chemical bonds, thereby possessing the ability to damage or destroy living cells.

Uranium-238, the most prevalent isotope in uranium ore, has a half-life of about 4.5 billion years; that is, half the atoms in any sample will decay in that amount of time. Uranium-238 decays by alpha emission into thorium-234, which itself decays by beta emission to protactinium-234, which decays by beta emission to uranium-234, and so on. The various decay products, (sometimes referred to as "progeny" or "daughters"), form a series starting at uranium-238,

TABLE 11. **Summary of uranium isotopes**

ISOTOPE	PERCENT IN NATURAL URANIUM	NO. OF PROTONS	NO. OF NEUTRONS	HALF-LIFE (IN YEARS)
Uranium-238	99.284	92	146	4.46 billion
Uranium-235	0.711	92	143	704 million
Uranium-234	0.0055	92	142	245,000

[*]Terms in bold are defined in glossary.

FIGURE 6. Uranium-238 decay series

Uranium-238
(half-life: 4.46 billion years)
⇩ alpha decay

Thorium-234
(half-life: 24.10 days)
⇩ beta decay

Protactinium-234m
(half-life: 1.17 minutes)
⇩ beta decay

Uranium-234
(half-life: 245,000 years)
⇩ alpha decay

Thorium-230
(half-life: 75,400 years)
⇩ alpha decay

Radium-226
(half-life: 1,600 years)
⇩ alpha decay

Radon-222
(half-life: 3.82 days)
⇩ alpha decay

Polonium-218
(half-life: 3.11 minutes)
⇩ alpha decay

Lead-214
(half-life: 26.8 minutes)
⇩ beta decay

Bismuth-214
(half-life: 19.9 minutes)
⇩ beta decay

Polonium-214
(half-life: 163 microseconds)
⇩ alpha decay

Lead-210
(half-life: 22.3 years)
⇩ beta decay

Bismuth-210
(half-life: 5.01 days)
⇩ beta decay

Polonium-210
(half-life: 138.4 days)
⇩ alpha decay

Lead-206
(stable)

Appendix C. Uranium: Its Uses and Hazards

as shown in Figure 6 (see previous page). After several more alpha and beta decays, the series ends with the stable isotope lead-206.

Uranium-238 emits alpha particles, which are less penetrating than other forms of radiation, and weak gamma rays. As long as it remains outside the body, uranium poses little health hazard (mainly from the gamma rays). If inhaled or ingested, however, its radioactivity poses increased risks of lung cancer and bone cancer. Uranium is also chemically toxic at high concentrations and can cause damage to internal organs, notably the kidneys. Animal studies suggest that uranium may affect reproduction, the developing fetus,[127] and increase the risk of leukemia and soft tissue cancers.[128]

The property of uranium important for nuclear weapons and nuclear power is its ability to *fission*, or split into two lighter fragments when bombarded with neutrons, releasing energy in the process. Of the naturally-occuring uranium isotopes, only uranium-235 can sustain a *chain reaction*—a reaction in which each fission produces enough neutrons to trigger another, so that the fission process is maintained without any external source of neutrons.[129] In contrast, uranium-238 cannot sustain a chain reaction, but it can be converted to plutonium-239, which can.[130] Plutonium-239, virtually non-existent in nature, was used in the first atomic bomb tested July 16, 1945 and the bomb dropped on Nagasaki on August 9, 1945.

The Mining and Milling Process

Traditionally, uranium has been extracted from open pit and underground mines. In the past decade, alternative techniques such as *in-situ leach* mining, in which solutions are injected into underground deposits to dissolve uranium, have become more widely used. Most mines in the U.S. have shut down and imports account for about three-fourths of the roughly 16 metric tons of refined uranium used domestically each year—Canada being the largest single supplier.[131]

---

127 Agency for Toxic Substances and Disease Registry, "ATSDR Public Health Statement: Uranium," Atlanta, December 1990.

128 Filippova, L. G., A. P. Nifatov, and E. R. Lyubchanski, *Some of the long-term sequelae of giving rats enriched uranium* (in Russian), *Radiobiologiya*, v. 18, n. 3, 1978, pp. 400–405. Translated in NTIS UB/D/120-03 (DOE-TR-4/9), National Technical Information Service, Springfield, Virginia.

129 Uranium-235 and plutonium-239 are called "fissile" isotopes—defined as materials that can be fissioned by low-energy (ideally zero-energy) neutrons.

130 Uranium-238 is converted to plutonium-239 by bombarding it with neutrons:
$$U\text{-}238 + \text{neutron} \rightarrow U\text{-}239$$
$$U\text{-}239 \rightarrow Np\text{-}239 + \text{beta particle (electron)}$$
$$Np\text{-}239 \rightarrow Pu\text{-}239 + \text{beta particle (electron)}$$

131 Energy Information Administration, *Uranium Purchases Report 1992*, DOE/EIA-0570(92), Washington, D.C., August 1993. The number of conventional mines operating in the U.S. has declined from a peak of hundreds to zero in 1993; seven "non-conventional" mining operations (e.g., in-situ leach) accounted for all domestic ore production for that year (NUEXCO, *NUEXCO Review: 1993 Annual*, Denver, 1994).

The milling (refining) process extracts uranium oxide (U_3O_8) from ore to form *yellowcake*, a yellow or brown powder that contains about 90 percent uranium oxide.[132] Conventional mining techniques generate a substantial quantity of **mill tailings** waste during the milling phase, because the usable portion is generally less than one percent of the ore. (In-situ leach mining leaves the unusable portion in the ground and does not generate this form of waste.) The total volume of mill tailings generated in the U.S. is over 95 percent of the volume of all radioactive waste from all stages of nuclear weapons and power production.[133] While the hazard per gram of mill tailings is low relative to most other radioactive wastes, the large volume and lack of regulations until 1980 have resulted in widespread environmental contamination. Moreover, the half-lives of the principal radioactive components of mill tailings, thorium-230 and radium-226, are long, being about 75,000 years and 1,600 years respectively.

The most serious health hazard associated with uranium mining is lung cancer due to inhaling uranium decay products. Uranium mill tailings contain radioactive materials, notably radium-226, and heavy metals (e.g., manganese and molybdenum) which can leach into groundwater. Near tailings piles, water samples have shown levels of some contaminants at hundreds of times the government's acceptable level for drinking water.[134]

Mining and milling operations in the U.S. have disproportionately affected indigenous populations around the globe. For example, nearly one third of all mill tailings from abandoned mill operations are on lands of the Navajo Nation alone. Many Native Americans have died of lung cancers linked to their work in uranium mines. Others continue to suffer the effects of land and water contamination due to seepage and spills from mill tailings piles.[135,136]

Conversion and Enrichment

Uranium is generally used in reactors in the form of uranium dioxide (UO_2) or uranium metal; nuclear weapons use the metallic form. Production of

---

132   Benedict, Manson, Thomas Pigford, and Hans Wolfgang Levi, *Nuclear Chemical Engineering*, 2d ed. (New York: McGraw-Hill Book Company, 1981), p. 265. Note that pure U_3O_8 is black. Yellowcake gets its color from the presence of ammonium diuranate.

133   Based on the total volume of all radioactive waste (including spent fuel, high-level waste, transuranic waste, low-level waste and uranium mill tailings) from all sources (both commercial and military) produced in the U.S. since the 1940s, as compiled in Scott Saleska, et al. *Nuclear Legacy: An Overview of the Places, Policies, and Problems of Radioactive Waste in the United States* (Washington, D.C.: Public Citizen, 1989), Appendix C.

134   U.S. Environmental Protection Agency, *Final Environmental Impact Statement for Standards for the Control of Byproduct Materials from Uranium Ore Processing*, Washington, D.C., 1983, v. 1, pp. D-12, D-13.

135   Gilles, Cate, Marti Reed, and Jacques Seronde, "Our Uranium Legacy," 1990 [available from Southwest Research and Information Center, Albuquerque, New Mexico].

136   In 1979, a dam holding water in a mill tailings settling pond at the United Nuclear Fuels Corporation mill near Church Rock, New Mexico gave way and released about 100 million gallons of contaminated water into the Puerco River which cuts through Navajo grazing lands.

Appendix C. Uranium: Its Uses and Hazards

uranium dioxide or metal requires chemical processing of yellowcake. Further, most civilian and many military reactors require uranium that has a higher proportion of uranium-235 than present in natural uranium. The process used to increase the amount of uranium-235 relative to uranium-238 is known as *uranium enrichment*.

U.S. civilian power plants typically use 3 to 5 percent uranium-235. Weapons use "highly enriched uranium" (HEU) with over 90 percent uranium-235. Some research reactors and all U.S. naval reactors also use HEU.

To enrich uranium, it must first be converted to the chemical form uranium hexafluoride (UF_6). After enrichment, UF_6 is chemically converted to uranium dioxide or metal. A major hazard in both the uranium conversion and uranium enrichment processes comes from the handling of uranium hexafluoride, which is chemically toxic as well as radioactive. Moreover, it reacts readily with moisture, releasing highly toxic hydrofluoric acid. Conversion and enrichment facilities have had a number of accidents involving uranium hexafluoride.[137]

The bulk of waste from the enrichment process is *depleted uranium*—so-called because most of the uranium-235 has been extracted from it. Depleted uranium has been used by the U.S. military to fabricate armor-piercing conventional weapons and tank armor plating. It was incorporated into these conventional weapons without informing operating personnel that it is a radioactive material and without procedures for measuring the doses to those personnel.

The enrichment process can also be reversed. Highly enriched uranium can be diluted, or "blended down" with depleted, natural, or very low enriched uranium to produce 3 to 5 percent low enriched reactor fuel. Uranium metal at various enrichments must be chemically processed so that it can be blended into a homogeneous material at one enrichment level. As a result, the health and environmental risks of blending are similar to those for uranium conversion and enrichment.

Regulations in the U.S.

In 1983 the federal government set standards for controlling pollution from active and abandoned mill tailings piles resulting from yellowcake production. The principal goals of these regulations are to limit the seepage of radionuclides and heavy metals into groundwater and reduce emissions of radon-222 to the air.

Mandatory standards for **decommissioning** nuclear facilities including conversion and enrichment facilities are only now being developed by the U.S. Environmental Protection Agency and the U.S. Nuclear Regulatory Commission (NRC). So far, the NRC has been using guidelines developed by its staff in 1981 to oversee decommissioning efforts.[138]

---

137 One such accident at the Sequoyah Fuels conversion plant in Gore, Oklahoma killed a worker and hospitalized 42 other workers and approximately 100 residents.

138 For more information about cleanup standards, see *Science for Democratic Action* (IEER, Takoma Park, MD), v. 3, n. 1, Winter 1994.

The Future

Uranium and associated decay products thorium-230 and radium-226 will remain hazardous for thousands of years. Current U.S. regulations, however, cover a period of 1,000 years for mill tailings and at most 500 years for **"low-level" radioactive waste**. This means that future generations—far beyond those promised protection by these regulations—will likely face significant risks from uranium mining, milling, and processing activities.

Superphénix nuclear reactor, France—it can create more plutonium or use it as a fuel. (Photo by Robert Del Tredici, Atomic Photographers Guild)

ABOUT THE AUTHORS

Arjun Makhijani has authored and co-authored numerous studies and books on nuclear-weapons-related issues. He is president of the Institute for Energy and Environmental Research and holds a Ph.D. in engineering from the University of California at Berkeley.

Annie Makhijani is Project Scientist at the Institute for Energy and Environmental Research and holds an M.S. in chemistry from the University of Maryland at College Park. She is responsible for research on radiochemistry-related issues at IEER. She has co-authored several reports on protection of the stratospheric ozone layer.